TRACTORS

JOHN DEERE

TRACTORS

AN ILLUSTRATED CHRONOLOGICAL HISTORY

Peter Henshaw

GREENWICH
EDITIONS

Published in 2005 by
Greenwich Editions
The Chrysalis Building
Bramley Road
London W10 6SP, United Kingdom

© Salamander Books 2003, 2004, 2005

An imprint of Chrysalis Books Group plc

All correspondence concerning the content of this
volume should be addressed to Salamander Books.

Designed and edited by:
FOCUS PUBLISHING, 11a St Botolph's Road,
Sevenoaks, Kent, England TN13 3AJ
Editor: Guy Croton
Designer: Philip Clucas MSIAD
Salamander editor: Marie Clayton
Photography by: Andrew Morland
"Classic Profile" photography by: Neil Sutherland

ISBN 0-86288-637-6
Printed in Malaysia

Contents

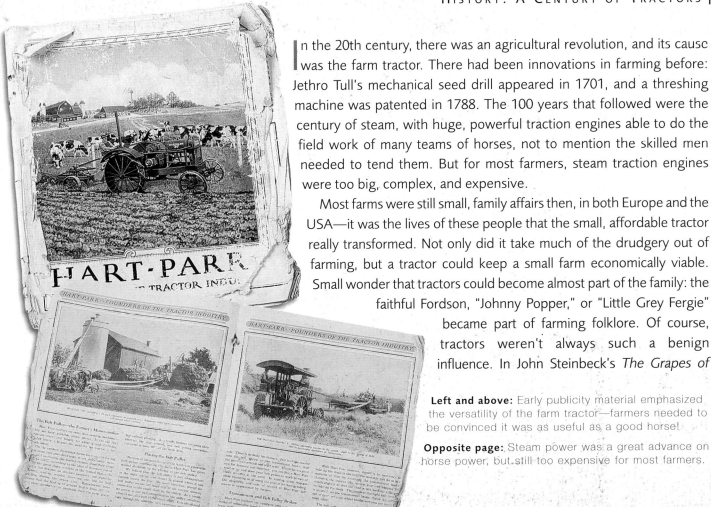

In the 20th century, there was an agricultural revolution, and its cause was the farm tractor. There had been innovations in farming before: Jethro Tull's mechanical seed drill appeared in 1701, and a threshing machine was patented in 1788. The 100 years that followed were the century of steam, with huge, powerful traction engines able to do the field work of many teams of horses, not to mention the skilled men needed to tend them. But for most farmers, steam traction engines were too big, complex, and expensive.

Most farms were still small, family affairs then, in both Europe and the USA—it was the lives of these people that the small, affordable tractor really transformed. Not only did it take much of the drudgery out of farming, but a tractor could keep a small farm economically viable. Small wonder that tractors could become almost part of the family: the faithful Fordson, "Johnny Popper," or "Little Grey Fergie" became part of farming folklore. Of course, tractors weren't always such a benign influence. In John Steinbeck's *The Grapes of*

Left and above: Early publicity material emphasized the versatility of the farm tractor—farmers needed to be convinced it was as useful as a good horse!

Opposite page: Steam power was a great advance on horse power, but still too expensive for most farmers.

Wrath, they are portrayed as the tool of big money, driving small farmers off the land—although thousands of farmers were kept in business by the ultra-affordable Fordson, some poor folks couldn't even afford one of those. But all that was to come.

Above and right: Waterloo Boy was one of the pioneers, appealing to traditionalists with this homespun image.

It was in the late 19th century when a number of European engineers perfected the internal combustion engine. It was a combination of things, notably the Otto four-stroke cycle and spark ignition, that made this power unit practical for the first time. More powerful, compact, and convenient than the big, heavy, oil or steam engines of the time, it offered the promise of mechanical power for everyone. Its impact was as profound as that of the silicon chip in the late 20th century, and the internet in the early 21st.

The Pioneers

As early as 1892, inventors were applying the new engine to tractors. John Froelich of Iowa built one powered by a Van Duzen engine, and formed the Waterloo Gasoline

Traction Engine Company. The Huber Company of Ohio used Van Duzen engines too, while Charles Hart and Charles Parr joined forces to build their first gasoline tractor in 1902. A year later, Englishman Dan Albone began building his three-wheeled Ivel tractor. It's interesting to note though, that few of these early tractor pioneers were new ventures. Most were established companies, either in the agricultural industry or in general engineering. John Deere, Case, Allis-Chalmers, Oliver, International,

Below: Alongside the new gasoline tractors, steam was still favored for heavy work by some.

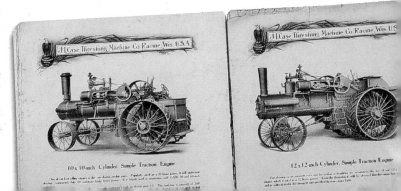

HENRY FORD

Henry Ford came from farming stock. Growing up on his father's farm in Dearborn, Michigan spurred the young Henry into an interest in engineering. It also instilled an understanding of the hard work faced by 19th century farmers—his development of a robust affordable tractor was perhaps inevitable.

In his thirties, Ford built his first car, a two-cylinder gasoline machine that could top 20mph, and after a few false starts set up in business to make the Model A in 1903. What Ford wanted, above all else, was to make a cheap car that lots of people could afford, and it worked. Within a few years, Ford was the world's biggest car maker. As Ford made more and more cars on a mass scale, they became cheaper and cheaper—success bred success. By 1915, a number of companies such as Pulford and Eros were already offering conversions to turn the Model T Ford into a basic tractor. That same year, Ford announced he would build a proper mass-produced two-plough tractor, at the unheard-of price of $200. Like the T, it would be simple, robust, and cheap. When it arrived, the Fordson Model F was all of that—and Henry's farming background had paid off.

Above: Henry Ford was experimenting with tractors long before the famous Fordson F finally went on sale in 1917. This is his "autoplow" of 1907, showing clear automobile influence.

Above: Advance-Rumely OilPull—fueled by kerosene, but still a big steam engine in its low-revving, easy-going nature, and later outmoded by lighter, cheaper gasoline machines.

Massey-Harris and Minneapolis-Moline—all famous names in tractor making, but all well established before they ever built a tractor.

At first, many of the early gasoline tractors were great, heavy things, as it was assumed they would replace existing steam traction engines, and do the same job. The Advance-Rumely OilPulls were typical, with massive, slow revving, twin-cylinder engines, such as the 654ci version fitted to the 16-30 of 1918. This turned at a mere

Left: International gear-driven Type A Mogul, sold by McCormick dealers. The modern IH logo shown here was added later, having been designed in the 1940s by renowned stylist Raymond Lowey.

530rpm but produced a useful 30hp at the belt, a respectable figure for the time. The OilPulls were not actually gasoline tractors at all, but ran on kerosene, relying on high cylinder temperatures (not spark ignition) to burn the fuel. With their chimney-like cooling towers puffing out kerosene smoke, the OilPulls could be mistaken for steam engines, and often were. International made a similar machine, sold as the Mogul, and more than 600 of its Type A found customers between 1907 and 1911.

But these tractors were still too big and expensive for the average farmer—to really expand the market, the manufacturers had to come up with something smaller and lighter. There were of course self-propelled two-wheel machines produced by Moline and Allis-Chalmers, but these weren't tractors proper—the Allis-Chalmers machine was designed to hitch up to existing horse-drawn ploughs and other implements with only minimal changes, but was not a success. But from about 1912, smaller tractors did begin to trickle onto the market. The Wallis was unveiled in 1913, bristling with innovation. Instead of a heavy separate chassis, it was of unit construction, while the engine was a car-style four-cylinder unit of 13/25hp. There were four wheels and from 1919 a fully-enclosed drive train. Case announced the 12/25 in 1913, while Happy Farmer, Farmer Boy, and Steel King also made small tractors. International joined in as well, with the little 8-16hp Titan Junior in 1917.

All these pioneers were American, and there was a good reason for that. In the USA, particularly in the Midwest, farms were larger and labor more scarce and expensive than in Europe. Here, the economic benefits of a tractor were immediately obvious, whereas in Europe

The NEW CA

Large Steering Wheel

Hardened Roller Chain Drive

Swinging Drawbar

TWO SIZES
MODEL "L"
PULLS A 3 OR 4
BOTTOM 14"
PLOW
MODEL "C"
PULLS A 2 OR 3
BOTTOM 14"
PLOW

Simple Ha

Above: Small tractors like this 10-20 Titan established International as a major manufacturer.

Left: By the 1920s, many elements of the modern tractor were in place, though there was plenty of room for improvement—chain drive (as on this Case) would soon be gone.

horse teams were still a common sight up to the Second World War. In fact, America enjoyed a tractor boom in the first two decades of the century—in 1906, there were only half a dozen tractor manufacturers, but by 1917, 260 were listed. Of course, few of these were to survive the depression, and by the 1930s the number of major American manufacturers was back to single figures.

If the truth be told, some of those manufacturers deserved to fall by the wayside. Faced with customers who often knew nothing about mechanical devices, they could make exaggerated claims which often had little connection with reality. A cartoon in Farm Implement News of 1917 showed a sharp-faced promoter for the "Skinum Tractor Company" gloating, fat cigar in mouth, over the huge profits of his crooked company. Fortunately, help was at hand.

Wilmot F Crozier, a former teacher, had gone back to farming, but found that the Bull and Ford tractors he bought were way below standard. So he sponsored a bill (passed in 1920 by the Nebraska State Legislature) which made testing mandatory for every model of tractor sold in the State. The University of Nebraska took on the job,

Above: Revolution (1)—Harry Ferguson's three-point hitch (here demonstrated on a Fordson in Norfolk, England) was a huge leap forward in tractor design, still used today.

Left: Revolution (2)—Henry Ford's Fordson F was, by the standards of its time, staggeringly cheap. Like the first home computers, it served to massively expand the market.

and drew up a detailed standard test, which measured things like horsepower, payload, and fuel consumption. So thorough and impartial were these tests, that they soon became a virtual national standard—they are still used by tractor buyers across the world.

The Fordson Revolution

In any case, the entire tractor industry already had a new standard to judge itself by—the Fordson F of 1917. This was one of the most significant tractors of all time; just as the Model T Ford had brought cheap motoring to the masses (or middle classes, at least) so the Fordson F did the same for farm tractors. Compared to rivals, it was simple, easy to understand, and above all, cheap. Henry Ford (he couldn't use his own name, as this was already in use by another tractor maker) had been experimenting

Below: It may not look streamlined now, but in 1917 Twin City's 16-30, with its full bodywork and large rear fenders, looked more like a car than any tractor.

with a "peoples' tractor" for ten years, and the first prototypes were little more than Model T's minus bodywork and with tractor-type steel wheels. But the production F was quite different—using the engine and transmission as stressed members did away with the chassis, making the tractor lighter and cheaper to produce. This in turn gave a better power to weight ratio, despite the relatively small 251ci engine.

Everything about the F was aimed at making it as cheap as possible—it was even made short enough to be parked sideways on a rail wagon, to minimize shipping costs. This attention to detail showed in the price of just $795 when it was launched, later reduced to an incredible $230. At these prices, the Fordson persuaded a whole generation of farmers to buy their first tractors, and over half a million were sold. Rivals had a choice of offering something more, cutting their prices to Ford

Above: Getting there...fully enclosed transmission, including the clutch, with gears and final drive all running in oil, was another big advance—far more long-lived than chain drive or exposed gears.

Left: "Twin City" tractors were built by Minneapolis Steel & Machinery—the name came from the location, midway between Minneapolis and St Paul, Minnesota. Merger with the Moline Pow Company in 1929 produced Minneapolis-Moline.

levels... or going out of business. Ironically, despite this success, Henry Ford chose to leave the American tractor business in 1927, to clear production space for his new Model A car, but the Fordson lived on, produced in England and Ireland.

JEROME CASE

Jerome Increase Case was a man of many parts. As well as being an inventor and manufacturer, notably of threshers and steam engines, he founded two banks in Wisconsin. He was president of the Agricultural societies for both Wisconsin and Racine County and founded the Wisconsin Academy of Science, Arts and Letters. As if that wasn't enough, he was mayor of Racine three times and served as a state senator.

In between all this, Case built horse-drawn threshing machines—so many of them, and so successfully, that he became known as the "Threshing Machine King." Case was a hands-on man—one story goes that in 1884, a faulty thresher could not be repaired by the local Minnesota dealer. The 65-year-old Case made the trip there himself, and when even he could not cure the fault, promptly burnt the rogue machine to the ground. He ordered that a brand new replacement be delivered the following day, and it was! By then, Case was also the biggest manufacturer of steam engines in the world, and entering the gasoline tractor market was a natural progression.

Above: Old Abe the eagle was the Case symbol for nearly 100 years. Named after President Abraham Lincoln, he was the mascot of Company C of the Eighth Wisconsin Regiment. J.I. Case was so impressed he adopted Abe as a company mascot as well. Old Abe wasn't dropped until 1969.

You might have expected Ford to have just eliminated the competition, but things weren't really that simple. The Fordson F was a small, and in some ways quite crude, machine, and there was plenty of demand for something more sophisticated. Not that the John Deere D could be accused of the latter. It never sold as well as the Fordson, but by any other standards was a huge success. Announced in 1923, it enjoyed a 30-year production run, and established the classic John Deere horizontal twin-cylinder layout—"Johnny Popper" was born. The secret, and one which ensured John Deere's success for the next 40 years, was one that placed rugged reliability above everything else—there were only two transmission speeds, and the twin-cylinder engine was a slow-turning plodder, but it certainly was reliable. By the time production of the John Deere D ended in 1953, over 160,000 had been sold.

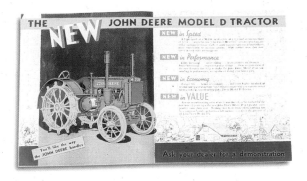

Above: Johnny Popper—Model D established the classic twin-cylinder John Deere layout
Left: Revolution (3)—Allis-Chalmers pioneered the use of pneumatic rubber tires on tractors. There were huge benefits in speed, efficiency, and comfort and another milestone in tractor design.

Until 1924, when the International Farmall arrived, tractors were either small and light for cultivation, like the Fordson, or heavier and more powerful for drawbar or belt work, like the John Deere—you couldn't have both. But the Farmall, as its name suggested, did just that. With 24hp at the belt it could drive a thresher, but was nimble, able to turn in its own length, and had the ground clearance to drive between rows of cotton or corn without damaging the plants. It was just what the market had been waiting for, and over 4,000 Farmalls were sold in 1926, and ten times that in 1930. International later followed up with smaller and larger versions of the same thing—this was one of the landmark tractors.

But even the Farmall was restricted in one way. Like every other tractor of its time, it ran on steel wheels, which restricted its road speed to a fast walking pace.

JOHN DEERE

John Deere was a blacksmith by trade, born in Rutland, Vermont, in 1804. After serving a four-year apprenticeship, he made his name with high quality polished hay forks and shovels. But a depression in Vermont forced the young blacksmith to seek his fortune further west. He settled in Grand Detour, Illinois, sent for his wife and children, and soon had a thriving business.

But there was a problem. Many of the pioneer farmers (we're talking pre-railroad days here) were considering giving up on Illinois and moving on. The Midwestern soil was fertile, but rich and heavy, and clung to the bottom of cast iron ploughs, which were designed for the lighter, sandier soils of New England. But Deere found the solution —a carefully shaped, highly polished plough of good quality steel would scour itself. It was a huge success, and within ten years he was making 1,000 ploughs a year. In 1918, Deere & Co bought the Waterloo Gasoline Engine Co, and thus entered the tractor business.

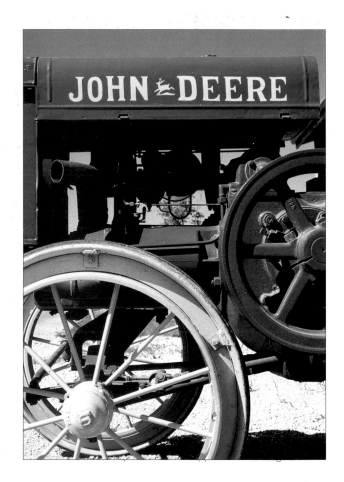

Right: Twin cylinders, rugged simplicity and complete reliability—they were the John Deere watchwords from 1923 right up to 1960. Few tractor mechanical layouts were so strongly associated with a single manufacturer.

Above: Revolution (4). The International 10-20 was competent enough, but it was the Farmall that really showed the way forward. Here at last was the do-it-all tractor, with the agility of a Fordson but the power to drive a thresher or do heavy work.

Pneumatic rubber tires would make for faster hauling on tarmac roads, be quieter, and give a more comfortable ride. Goodrich and Firestone had experimented with tractor pneumatics, but it was Allis-Chalmers which put the concept into practice, in 1929 on the U model. The effect was dramatic. The low pressure tires halved the tractor's power needs in the field, and allowed a high-speed fourth gear to be added to the transmission—15mph doesn't sound special now, but it was stratospheric in 1929. To underline the point, A-C hired a team of racing drivers and gave high-speed demonstration races at County Fairs, on specially geared-up U's. One even reached 68mph on the Utah salt flats! Rubber tires weren't cheap, at an extra $150, but the benefits were undeniable, and within a few years every major manufacturer was offering them.

The Ferguson Connection

Slowly, all the elements of the modern tractor were falling into place, and the single most significant one was about to be unveiled—the Ferguson three-point hitch. Until the early

Above: A marriage made in heaven—at least for a while. Harry Ferguson and Henry Ford (both far left) collaborated on the Ford 9N tractor, which combined Ford's mass production knowhow with Ferguson's three-point hitch.

Right: Oliver 70 brought six-cylinder sophistication to mid-range row-crop tractors.

1930s, hitching up implements was a laborious process, but the Ferguson system eliminated all that— you just reversed up, hooked on, and drove off. First using springs and levers, later hydraulics, this offered many advantages. Clever geometry meant that part of the implement's drag was used to exert downforce on the tractor's rear wheels, increasing traction; it had built in draft control; and finally, it would prevent the tractor tipping over backward, a useful safety feature in those days before safety cabs. In short, it was a huge leap forward, and today almost every tractor on the market has a three-point hitch. Harry Ferguson even designed a tractor to go round it, though he later entered into an agreement with Henry Ford with the 9N of 1939, which combined Ford's production knowhow with

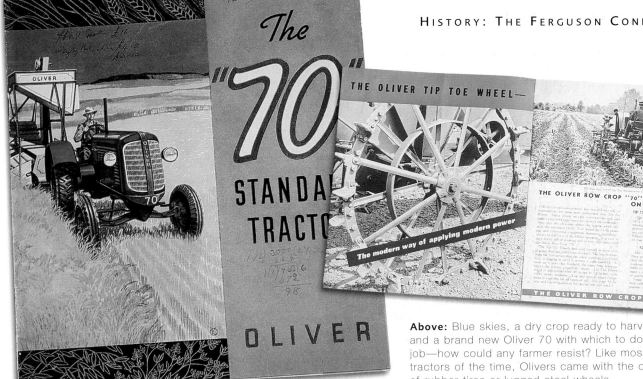

Above: Blue skies, a dry crop ready to harvest and a brand new Oliver 70 with which to do the job—how could any farmer resist? Like most tractors of the time, Olivers came with the choice of rubber tires or lugged steel wheels.

Ferguson's inventive genius. More than anyone else, these two men were co-fathers of the modern tractor.

The Ford 9N was a small tractor in the tradition of the original Fordson F, but some farmers were clamoring for more power. Oliver responded with the six-cylinder 70 of 1935. It didn't set any new power records, but six cylinders where two or four were the norm gave it unprecedented smoothness—it was a high revving, high compression engine that hinted at the power race to come. What was surprising was that this super-smooth

Above and right: All the comforts of home—or a well appointed car at least—while ploughing. With its fully-glazed cab (the first on a tractor), heater, and 40mph top speed, the Minneapolis Moline UDLX was, in 1938, 30 years ahead of its time. But having just survived the Depression, not many farmers were tempted by its high price.

tractor came from Oliver, which had produced few innovations apart from the splined rear axle, which allowed for easy adjustment of the rear track to suit different crop spacings. But if the Oliver 70 was a pointer to the future, the Minneapolis-Moline "Comfortractor" was ahead of its time. It was the first machine with a standard glazed cab. Not only that, but it boasted all the comforts of a high-class car – weather protection, heater, radio, and full instrumentation. It could also top 40mph on the road, and in theory one could use it for ploughing all day, then dress up and drive into town for the evening. But for 1938, it was just a bit too advanced for most farmers, and only 150 were sold.

The Second World War caused a hiatus in tractor

HARRY FERGUSON

Like Henry Ford, Harry Ferguson was the son of a farmer, though this farm was in Ireland, not the USA. Also like Henry, he could be difficult and irascible; more to the point, there was another point of similarity between the two men—both were immensely gifted engineers, and between them they changed the face of the farm tractor.

But unlike Henry Ford, Ferguson showed no early burning desire to work on agricultural machines. In his twenties, he concentrated instead on flying: he designed, built and flew a series of aeroplanes himself, and it wasn't until he became the Belfast agent for Overtime that he began to turn his attention to tractors. His implement hitch—first as a conversion for Ford Model Ts, later as the famous three-point hitch—was such an inspired piece of design that today it's a standard feature on almost every modern tractor.

Ferguson was a driven businessman as well as an engineer, and founded his own tractor company after agreements with Ford and David Brown broke down. He sold this to Massey-Harris in 1953—which resulted in Massey-Ferguson—and went on to design four-wheel-drive systems for cars. He died in 1960.

Above: A meeting of minds: Harry Ferguson demonstrates his three-point hitch to a highly receptive Henry Ford. The two men had similar backgrounds, but their gentleman's agreement, based on a handshake, was to end in court.

Left and above: Crawlers (notably Best and Caterpillar) played a minor role in agricultural work, but were still useful where traction or impaction was a problem for wheeled machines.

development, in North America as well as Europe. Tractors still rolled off the production lines, though in far smaller numbers than before—the manufacturers were often busy with military contracts. Massey-Harris built tanks; International Harvester and White made half-tracks; John Deere produced all sorts of things, including mobile laundry units. So tractor production went on, but some changes had to be made: a rubber shortage put an end to the pneumatic tire option for a while, while the Ford 9N became the wartime austerity 2N, with no rubber tires, battery/generator electrics, or electric start. The war put all tractor development on hold.

Playing Catch-Up

In the ten years that followed 1945, the tractor industry made up for lost time. The 1920s had been a time of innovation, but much of the '30s was taken up in recovering from the depression. Then the war delayed any further technical advance. Now the restraints were off, and the 1950s and '60s saw diesel engines, multi-speed transmissions and power figures that increased year on year.

Diesel power came first. This had been around for a long time—Rudolf Diesel had built the first practical engine that bore his name in 1907. Using compression ignition in place of an electrical spark, the diesel engine used less fuel than a gasoline equivalent and was more robust and reliable. In short, it was ideal for tractors, and Caterpillar had become a major user of diesel engines in its crawlers in the 1930s, some of which were put to work in the field.

CYRUS McCORMICK

Cyrus McCormick isn't one of those instantly recognizable names among tractor manufacturers, but without him there would have been no International Harvester. Robert McCormick had attempted to produce an improved reaper, but it was the young Cyrus —his son—who made it work in 1831 and patented the idea three years later. The McCormicks lived in Virginia, but like many before him, Cyrus realized that there were far greater markets in the Midwest, where huge prairies were being put to the plough for the first time.

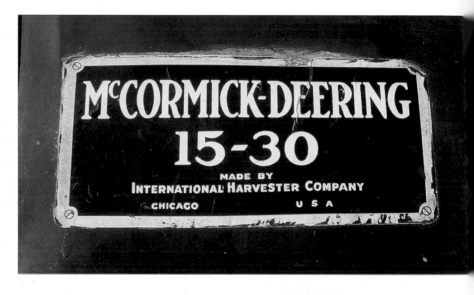

In 1847, McCormick started making reapers in Chicago, and despite several setbacks—the death of his brother and business partner in 1865, and the destruction of the factory by fire in 1871—survived and prospered. Cyrus, still at the helm, died in 1884, but his son Cyrus Jr carried on where he left off. In 1902, the McCormick Harvesting Machine Co merged with the Deering Harvester Co, and International Harvester—soon to be a major force in the tractor industry—was born.

Above: When bitter rivals McCormick and Deering merged in 1902 to form International Harvester, it was a good move for survival, but the rivalry went on for years. McCormick dealers refused to sell rebadged Deering tractors, and vice versa, so two model lines had to be developed. It took anti-trust action by the Justice Department to persuade IHC to consolidate its dealers under one name.

John Deere was first with a wheeled diesel tractor, in 1947—the R model had actually been on the drawing board during the war years. It was a diesel version of the classic John Deere twin-cylinder engine, the biggest JD yet, at 416ci (6.5 litres), not to mention the most powerful, with 51 PTO hp. But being a diesel, it was more fuel-efficient than any gasoline rival as well, and so cheaper to run—20,000 were built in three years. Other manufacturers soon followed John Deere with their own

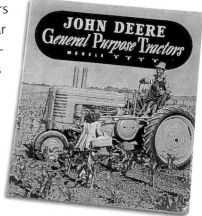

diesels. Allis-Chalmers announced the WD45 a year later, though the six-cylinder engine was bought-in from Buda of Illinois—it would be a few years before all the major manufacturers produced their own diesel engines.

Another example of this was the Oliver Super 99, which had the option of a General Motors three-cylinder two-stroke diesel. With super-charging and a 17:1 compression, this produced over 70hp, making the 99 Diesel the most powerful tractor of its time. It certainly had different power characteristics,

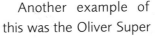

Left and above: Despite having stuck to the same engine layout for over 20 years, John Deere pioneered the model diesel tractor. The R (left) used a bigger diesel version of the existing twin-cylinder engine.

as the GM diesel needed to be revved high to keep that 70 horsepower coming. But it still came out more efficient than a gasoline engine of equivalent power—within ten years, diesel had become the standard tractor power unit, though some manufacturers continued to offer gasoline (especially in the smaller tractors) up to the early 1970s.

But the '50s saw more than just diesel power—a whole raft of new features appeared, as each manufacturer tried to outdo the others in this increasingly competitive market. Take the Allis WD45. As well as the new diesel option, it featured Power Shift, which allowed easy adjustment of rear wheel track. Instead of laboriously knocking the wheels into place, the driver now used engine power to slide the wheels in or out on spiral rails — this simple but effective system is still in use today. Two-Clutch Power Control allowed continuous power take-off, whether the tractor was moving or not, and the oil-bath transmission clutch stood up to deliberate slipping. Live hydraulics became standard fare, as did automatic draft

Right: Things were simpler in those days. A Farmall MTA's electrical box houses an ammeter, ignition and lighting switches, and a solitary fuse.

Below: A first for Farmall. The MTA was the first tractor to use a shift-on-the-move transmission, with a two-speed planetary gearbox giving ten forward ratios in all.

"Farmall" was such a strong brand, it still got biggest billing in the mid-'50s. McCormick survived as a maker's name, but International would eventually replace both.

control, which lifted the implement slightly in heavier patches of soil, to prevent the tractor bogging down.

But it was tractor transmissions that showed the single most important advance. Up until the Second World War, tractors used conventional single-range transmissions with three, four, or sometimes five speeds. But as tractors were expected to do a wider variety of jobs, it became more crucial to have exactly the right gear ratio for each one. That remains true today—a tractor's transmission design is often of more interest to farmers than its engine, and many of the recent innovations have centered around this feature.

In the 1950s, the big leap forward was to add a two-speed planetary gearbox to the main one, thus doubling the number of effective ratios to eight or ten. This allowed more precise control of speed in the field, as well as adding higher ratios for more speed on the road. International's Farmall MTA was one of the first with this feature, and its Torque Amplifier could be shifted between ranges without stopping the tractor. Within a few years, every major manufacturer was offering their own version of the same thing, and the marketing men

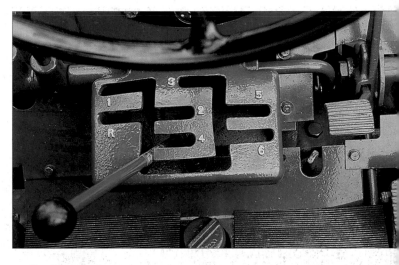

Above: Multi-speed transmissions arrived in the 1950s, first with the addition of a simple two-speed gearbox to the main one. The number of ratios, and complexity, increased in succeeding years.
Far left: Two-stroke GM diesel made the Oliver Super 99 the most powerful tractor of its time.

came up with suitably inspiring names. Allis-Chalmers' version was the "Power Director," Minneapolis-Moline favored the "Ampli-Torc," while International Harvester went for the more straight laced "Torque Amplifier." Case

You'll handle more acres per day with CASE HI-TORQUE POWER

Two new *Comfort King* Tractors

Left: Into the '60s with Case-O-Matic, which allied a torque converter with an 8-speed transmission.

Below: Eventually, tractor drivers got a better deal. This Comfort King Case was an early factory-built cab.

simplest answer was to increase engine size, but this also made engines heavier and more cumbersome, and could bring vibration problems. The answer—and another landmark in tractor design—was turbocharging.

had the "Case-O-Matic," which used a torque converter.

The Power Race

As the 1950s gave way to the '60s, it seemed as if the whole world had gone speed and power crazy. Airliners, cars, motorcycles, and trucks—all sought the same thing. Tractors were no exception, though farmers could point out that 20% more power could mean 25% more work in the field. The

at the new 5020
you

Allis-Chalmers' D19 of 1961 was the first production turbocharged tractor. Turbocharging A-C's 262ci (4.1 liter) four-cylinder diesel gave 25% more power, to 67PTO hp and 62hp at the drawbar—not much by today's standards, but respectable enough for the early 1960s. For the first time, farmers could have gasoline power with diesel economy. The D19 and its turbo descendents became favorites in the sport of tractor pulling.

But the D19 wasn't a front runner in the power race for very long. By the middle of the decade, there was a choice of several 100hp machines. Allis itself designed an all-new

Above and right: All change for John Deere in 1960, when it replaced the old twin cylinder tractors with a "New Generation" of four- and six-cylinder machines. They were up to the minute, with full hydraulics, quick-attach couplings, and improved ergonomics. "Johnny Popper" was put out to pasture...

Eleven models – 34 to 143 horsepower...

New 820
34 horse power—
tough, compact, ready to
tackle any one of a hundred
different jobs.

New 920
40 horse power—
ideal for small and medium
acreage farms, or as a
second tractor on big-scale
operations.

New 1020 Series
47 horse power—
compact, stable, versatile.
Available in Utility, Vineyard
and Orchard models.

New 1120
52 horse power—
powerful all-round tractor,
ideal both for cultivating
and all kinds of utility jobs.

New 2020 Series
64 horse power—for ample
power and steady reliable
service year in, year out. Available
in Agricultural and Orchard models.

Test a new
John Deere tractor
on your farm.
Just contact your
John Deere dealer

Above: Chevron-tread front tires denote four-wheel drive on this 1977 John Deere 4040S.

Right: John Deere's Sound Gard cab, another big step in cab design, featuring built in roll-over protection and optional air conditioning.

direct injection diesel of 426ci (6.6 liter) for the 103hp D21. This was in a different league to earlier tractors, and the D21 needed a new set of implements to suit its high-power capabilities, including a seven-bottom plough. Case came up with its most powerful tractor yet, increasing engine size to 451ci (7.0 liter) to make one of the first 100hp machines. Not to be left behind, International Harvester turbocharged its existing D361 six-cylinder diesel to achieve the same result. One interesting aside was that this tractor, the 1206, had to have a beefed up transmission, with hardened gears, heavier pinions, and final drive gears, to cope with the extra power. Even the tires had to be specially designed—the 1206 had so much power that conventional tires buckled their sidewalls or simply span off the rim.

For some farmers, even these power increases weren't enough, and for their needs the 1960s and '70s saw the development of the giant super-tractor. These were designed for the massive wheat fields of the Midwest, where farms were big enough to make such massive machines viable. The Steiger brothers farmed 4,000 acres in Minnesota, and couldn't find a tractor large enough for their needs, so they built one. It set the pattern for all subsequent Steigers: four-wheel-drive, articulated drive shaft, and a big, powerful diesel engine. It was so successful, they went into business, and throughout the '60s and '70s built super-tractors both for themselves and for the major manufacturers—Ford and Allis-Chalmers both sold Steigers under their own badges. Typical was the ST325 Panther (the Steigers loved to name their tractors after big cats) with 270hp Cummins diesel and an all-up weight of over 15 tons. Steiger was taken over by Case IH in 1986, but its legacy—not to mention the name—lives on. Big Bud was another super-tractor of the time, offering over 500hp machines by 1977.

However, as the power race forged on, the driver wasn't being forgotten—at least from the 1970s, which saw the increasing use of cabs, often with ROPS (roll-

Above: Giant four-wheel-drive super-tractors became increasingly common in the 1970s and '80s. This is the Case-Steiger range.

over safety protection) and a greater emphasis on driver comfort. There were good reasons for this—not only was a comfortable driver able to work longer and make fewer mistakes, but many tractors were still bought by owner-drivers! John Deere had actually developed a ROPS cab back in 1966, and took the unusual step of releasing its patents to the entire industry, to encourage other manufacturers to follow suit. In time, ROPS became a legal requirement, and saved many lives.

There were moves to cut down noise levels as well—

Above: A new name for tractors. Truck maker White bought Oliver, Minneapolis-Moline and Cockshutt in the early 1960s.

Left: The new generation of big tractors weren't cheap, but paid for themselves in doing more work in less time.

the Nebraska tests included noise levels from 1970, and John Deere was at the forefront of this, with the "Generation II" 30-series tractors of 1972. Their Sound-Gard cab, with its large curved windshield and tinted glass, was quiet enough to allow the option of a radio/cassette. It had built-in ROPS and the option of a pressurizer to keep dust out, or air conditioning. Over half of 30-series buyers paid extra for the new cab. International's Pro Ag line of 1976 showed similar advances, with a big 43.2 square feet of glass area. The seat was placed forward of the rear axle, for a smoother ride and to ease access, and the doors were of double-wall construction, with heavy rubber seals to keep out dust. Air conditioning was standard, with radio, cassette, and hydraulically mounted seat all on the options list.

More New Ideas

Meanwhile, there were more drivetrain advances. Things had moved on since the 1950s, and Allis-Chalmers's 8-speed Power Director had been eclipsed by 12- and 16-speed rivals. Allis responded with a new 20-speed version in the new 7000-series machines, while Case offered its 12-speeder in the 970. This used a combination of four mechanical gear ratios and three powershift ranges. Powershift, as the name suggested, could change gear on the move, under power, though there was an interlock to prevent mistaken forward/reverse selection. As for power, if a

Above: Big articulated four-wheel-drives from Case, now including the David Brown name.

chunky the tires, and all-wheel-drive became an option right down to some of the smallest models. At the other end of the range, most manufacturers were offering giant super-tractors, just one class down from the Steiger size. All used large turbocharged diesels (often bought in from Cummins or Caterpillar), and four-wheel-drive: examples were the Case Magnum, which offered 155-264hp and is still part of the Case line-up. John Deere went the whole hog with its 8850, designing its own intercooled 300hp V8 diesel to suit, plus a 16-speed transmission.

There were some genuinely new concepts as well. JCB, a British maker of mechanical excavators, jumped into the tractor market with the Fastrac. This offered conventional field capabilities and a road speed of up to 45mph. It was popular with farmers who had to cover distance between field and farm, and soon inspired a number of similar rivals. Caterpillar had been on the fringes of the tractor market for decades, but was a

turbocharged diesel wasn't enough, most manufacturers were now offering intercooling as well—the Allis 7000 came in 130hp turbo and 156hp intercooled forms.

Into the 1980s, four-wheel-drive also became commonplace. There was after all a limit to how much power could be fed through two rear wheels, however

Far right: Now that's big. As power outputs increased (this Case International 9280 has 375bhp) traction problems did, too—triple tire sets were one solution.

Above: A 22,000lb Ford 946, a far cry from the original Fordson F. Note the use of the Versatile name—this had started out in 1967 as an independent maker of big 4x4 tractors, but was taken over by Ford-New Holland in 1987.

new entrant with its Challenger in the early 1990s. This used rubber tracks, far more suited to boggy fields than the traditional steel tracks, and faster and quieter

Remember when you first decided to farm?

on tarmac as well. Again, it spawned rivals.

While all this was going on, the world tractor industry was in turmoil. A worldwide recession in the 1980s hit hard, while in the '90s the opening up of the former USSR and Eastern Europe brought fresh challenges. For a while, it looked as if the USA might pull out of tractor manufacturing altogether, but John Deere came up with a new range of US-built mid-range machines. For most, merger and take over was the only way to survive. Case swallowed International, and Allis-Chalmers closed down, only to be bought up by AGCO. The latter became a real success story, buying other famous names like White and Massey-Ferguson, and keeping them alive.

Into the 1990s, it looked as if sheer power was no longer the ultimate goal of tractor design. Instead, the tendency was to use electronics and "intelligent"

Above and above left: New life for Allis-Chalmers in the 1990s, first as Deutz-Allis (left) then under the AGCO umbrella (above). The traditional Persian Orange remained, to appeal to folk-memories of grandad's Model U.

transmissions to make better use of what power there was. In 2002, fully automatic "stepless" transmissions were introduced. Electronics in particular, offered the promise of more precise work than even the most skilled driver. In theory, all he or she had to do was punch in the parameters, and the tractor's electronic "brain" would do the rest, controlling speed, draft, and revs to keep the tractor working as efficiently as

possible. In practice, these systems were and are very expensive, and some suffered teething troubles, but there's no doubt that in tractor manufacturing electronics are here to stay. We've come a long way since the Fordson F.

Above: Massey-Ferguson, White, and Allis—now all part of the AGCO empire, which has factories all over the world and a turnover in 2001 of $2.5 billion.

Right: Will bigger always be better? The latest trends suggest more efficient use of existing power is on the way, rather than more brute force.

Allis-Chalmers

Left: A stalwart of the industry, Allis-Chalmers always offered a range of competitive tractors.

Allis-Chalmers' 20-35 is a landmark tractor for two very good reasons. Along with the 18-30, it was the company's first really successful machine. Not only that, but it was a fine example of the galvanizing effect of the cut-price Fordson... But it took A-C several years, and plenty of dead ends, before getting this far. Four early attempts at building tractors, including a monstrous half-track and the little two-wheeled 6-12, failed to set the market alight. But in 1918, they finally made it with the 15-30—heavier and more expensive than a

SPECIFICATIONS

Engine	461ci (7.2 liter)
Transmission	2 speed
Horsepower	24/39hp
Weight	6,640lb (3,018kg)
Numbers built	n/a
Wheels & tires	n/a
Price new	$970 (1934)

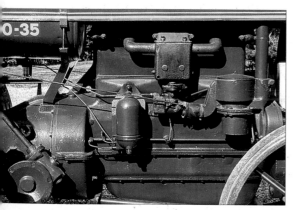

Above: A big, simple 461ci four-cylinder engine used a 4³⁄₄in bore and 6¹⁄₂in stroke to produce 39hp at the PTO.

Right: 1926 20-35 Special, with higher compression, sun canopy, and four-note whistle. This was the last of the expensive "long fender" 20-35s.

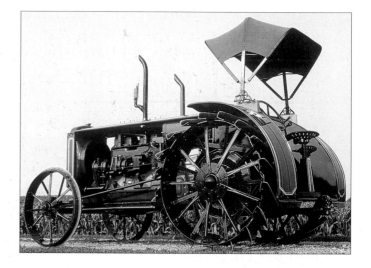

Right: Operating instructions for the 20-35: "Keep oil level to upper pet cock on left side of motor oil sump. Use oil of proper weight for season operating. Pressure gauge should indicate 15 pounds or more. Use grease gun on all fittings daily."

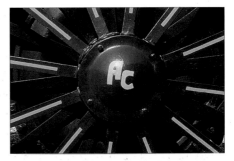

Above: A-C—Allis-Chalmers, a name that would become one of the big players in the American tractor business.

Left: This was Allis-Chalmers' big tractor of the 1920s, the 20-35. First listed in 1922, it was really no more than an uprated 18-30, and could be traced back to the 15-30 of 1918. Using A-C's own four-cylinder 461ci gasoline engine adjusted to give a higher running speed, an 18-30 was tested at Nebraska, giving 25hp at the drawbar, 44hp at the belt—the 20-35 was born. It was part of the range for over a decade.

Classic Profile

Left: 20-35 started out as a 15-30, but a higher rated speed brought a substantial power increase in 1922.

Right: Slashing the price of the 20-35 boosted sales to 4,800 by 1928. Henry Ford wasn't exactly quaking in his boots, but it was enough to give Allis-Chalmers a foothold in the tractor market.

Fordson, but more of a realistic sales proposition than previous A-Cs. The 15-30 was quickly renamed 18-30 when Nebraska tests revealed its true power output. Another test in 1922, using a higher rated speed from the 461ci four-cylinder engine, produced 25hp at the belt: the 20-35 was born.

But Allis-Chalmers hadn't struck gold just yet. The 20-35 was solid and reliable, but priced at over $2,000 it was still too expensive to sell in big numbers—less than 700 were sold in 1925, by which time Henry Ford had shifted over a million Fordsons. Undaunted, A-C's general manager Harry Merritt kickstarted a radical cost cutting programme.

The effect was dramatic, with the 20-35's price cut to $1,885 in 1926. The following year it was slashed to $1,495, and another $200 was lopped off the year after that. By 1934, the 20-35 cost a mere $970, brand new— less than half the original figure!

Sales took off—the 20-35 was the machine that established Allis-Chalmers in the tractor business.

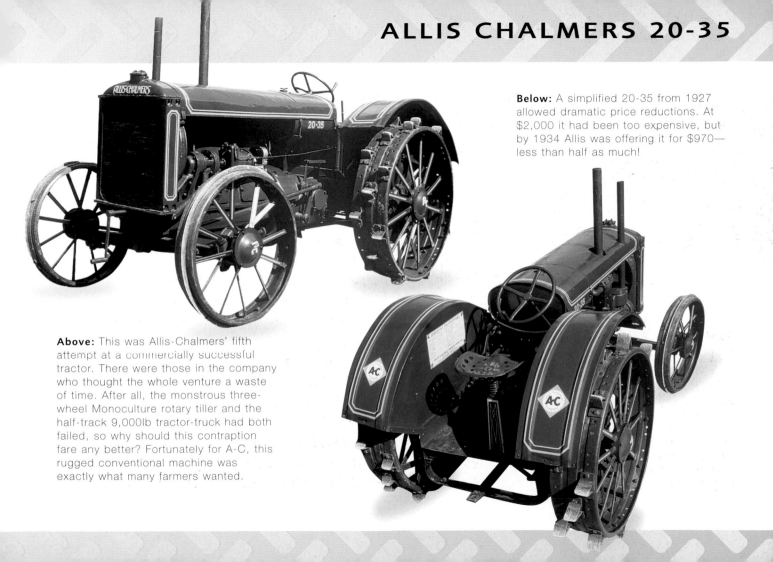

ALLIS CHALMERS 20-35

Below: A simplified 20-35 from 1927 allowed dramatic price reductions. At $2,000 it had been too expensive, but by 1934 Allis was offering it for $970— less than half as much!

Above: This was Allis-Chalmers' fifth attempt at a commercially successful tractor. There were those in the company who thought the whole venture a waste of time. After all, the monstrous three-wheel Monoculture rotary tiller and the half-track 9,000lb tractor-truck had both failed, so why should this contraption fare any better? Fortunately for A-C, this rugged conventional machine was exactly what many farmers wanted.

Opposite: Breakthrough! The Allis-Chalmers U was the first production tractor with pneumatic tires, which were a huge leap forward in tractor technology. They halved power requirements in the field, allowed higher road speeds, and were far more comfortable for the driver. Not a cheap option, but the benefits were undeniable —within a couple of years, every major manufacturer was offering pneumatics.

Left: There was something else new about the Allis U—the color. Until then, all A-Cs had been painted dark green. According to legend, the general manager was so taken with the color of some Californian wild poppies that he took a bunch back to the factory and decreed that this was to be the new corporate color scheme. Others say "Persian Orange" was simply the official color for United Tractors & Farm Equipment, for whom Allis originally designed the U. Take your pick!

0

Right: The WC was a huge success for Allis-Chalmers—29,000 were sold in its best year, and it remained part of the range up to 1948.

Left: Rubber-equipped WC. The drive-in attachment for implements made it easier to use than some rivals.

Right: By the end of the 1920s, Allis-Chalmers still didn't have a genuinely mass-market tractor—the WC of 1933 fitted the bill. Not an adventurous design, but light weight to make the most of its 21hp and, crucially, a pneumatic tire option—which no rival offered at the time—made the WC far more efficient than any comparable machine. The engine was a 202ci four-cylinder unit, notable for having the same bore and stroke dimensions.

Left: Caterpillar didn't have it all their own way, and some tractor makers offered crawler versions of their wheeled machines. This was the Allis-Chalmers M—basically the Model U with tracks and (understandably), minus the pneumatic tire option.

Left: Using the same 301ci engine as the U tractor rationalized components and reduced costs, not to mention producing 31.6hp at the belt. It was lower geared than the U, and able to pull 73% of its own weight in low gear.

Above: There were no frills on the M crawler, which did see some agricultural use as well as the more obvious construction markets. No diesel option yet (though Caterpillar already had this). Top speed was 4.2mph.

Left: Old before its time. The Model A Allis-Chalmers was a direct replacement for the venerable 20-35, and even used the same engine, now producing 33hp at the drawbar, 44hp at the belt (later boosted to 40/50hp). A new four-speed gearbox replaced the two-speed version.

Below: Pneumatic tires were Allis-Chalmers' big selling point in the 1930s—an expensive option, but one that paid dividends on hard worked tractors. They couldn't prevent the A from obsolescence, but helped the younger machines, such as this WF, a standard tread version of the WC row-crop.

Left: Announced in 1948 to replace the WC, A-C's WD used the same engine (albeit uprated) along with a lot of new features.

Right: Rear wheel track could be adjusted by engine power on the WD, which saved some hard work with a sledgehammer.

Below: The WD was later updated as the WD45, with 25% more power from a bigger engine. A diesel powered WD45D and in-seat hitching from the Snap Coupler were also features.

Above: Most harvesters still needed towing, though self-powered combines were on the way.

Classic Profile

ALLIS-CHALMERS

This, the Allis-Chalmers Model B, was one of the Milwaukee company's landmark tractors. In the mid 1930s, despite the efforts of Henry Ford, many small farms still hadn't bought a tractor—it wasn't always a matter of money, as often even a little Fordson could be too big for a really small holding. In 1937, Allis unveiled a new tractor aimed directly at small farmers, the B.

Weighing in at just 2,000lb, the B was lighter than just about anything else available. That enabled it to make best use of its modest 116ci power unit, as well as making the B very easy to handle, a fact underlined by its small dimensions. Best of all, it cost only $495, cheaper than any rival. The B was a huge success, and over 11,000 units were sold in the first year. By the time production ended in 1957, over 100,000 had been sold.

But how did Allis-Chalmers make the B so light weight? Put simply, there was no

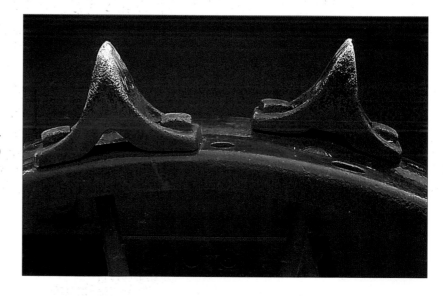

Below: Detail of the B's lugged steel wheels, fitted during the World War Two rubber shortage. These wheels could be widened by bolting on extensions, for greater traction. The lugs helped the traction as well, but made for slow progress on the road.

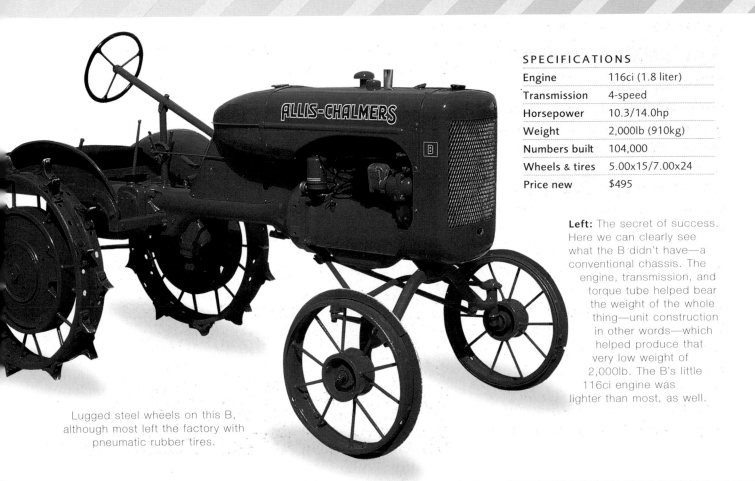

SPECIFICATIONS

Engine	116ci (1.8 liter)
Transmission	4-speed
Horsepower	10.3/14.0hp
Weight	2,000lb (910kg)
Numbers built	104,000
Wheels & tires	5.00x15/7.00x24
Price new	$495

Left: The secret of success. Here we can clearly see what the B didn't have—a conventional chassis. The engine, transmission, and torque tube helped bear the weight of the whole thing—unit construction in other words—which helped produce that very low weight of 2,000lb. The B's little 116ci engine was lighter than most, as well.

Lugged steel wheels on this B, although most left the factory with pneumatic rubber tires.

Above: The Allis-Chalmers B succeeded where others had failed, in persuading very small farmers to buy their first tractor. It had a low price, low weight, and easy handling.

chassis. Instead, the engine, transmission and torque tube were all used as stressed members, helping to support the whole assembly, and doing away with the need for a heavy separate chassis. It also gave the B its elegant wasp-waisted look, which was capitalized upon by industrial designer Brooks Stevens, and improved visibility without the need to offset the engine, thus upsetting stability.

Although a small, basic machine, the B could be upgraded with various options. It was designed from the start for pneumatic tires (and only the wartime Bs left the line without them) and by 1940 the price included electric lights and starting. You had to pay extra for a PTO shaft and pulley ($35), and an adjustable-width front axle (38 to 60 inches for only $20). There were power upgrades too. The first batch of 96 tractors used a bought-in 113ci Waukesha engine, while A-C was developing its own 116ci unit. That was later boosted to 125ci, improving on the 116's 10.3hp at the drawbar. That such low power figures produced such a useful tractor is testament to its lightweight design.

How much simpler
could a tractor
get?

Above: A slim waist gave the B
good visibility without resorting to
an offset power unit.

Above: Remote oil pressure gauge on this B, which comes
(like they all did) in traditional Allis-Chalmers Persian Orange.

Below: Rear-engined radical. The Allis-Chalmers Model G, unveiled in 1948, mounted its little 62ci Continental four-cylinder engine at the rear, to maximize traction. It was aimed at nurseries and market gardens, for whom small size and good visibility of the work in hand was paramount.

Right: Although the cheapest Allis at the time, the G came as standard with a four-speed gearbox and track adjustable between 36 and 64in—a belt pulley and hydraulic lift were optional extras.

Above: Allis-Chalmers WD45 of 1955 with side-delivery rake. Two-Clutch Power Control allowed continuous power take-off, whether the tractor was moving or not.

Right: Little D10 replaced the Model B in 1959, powered by a 139ci version of the bigger D14 power unit. D10 was the single-row tractor; the similar D12 straddled two rows.

Below: D15 was A-C's first move in the 1960s power race, with 18% more power than the D14 it replaced. Increased compression ratio and a 2,000rpm rated speed did the trick.

Right: D12 was a miniature version of the larger A-Cs, with Power Shift rear wheels, Traction Booster, and an independent PTO all available. There was no diesel option, though.

Left: D272 was the English Allis-Chalmers. A-C's factory in England had already made its own update on the Model B, with a diesel option—the D272 was another step forward.

Above: Traction Booster was another Allis feature of the 1950s. A hydraulic pump would lift the implement slightly when conditions got sticky, to prevent bogging down.

Today, turbocharging is commonplace—as likely to be found on a standard family saloon as anything more exotic. Among tractors, a turbo-diesel is the most common form of power unit—but it wasn't always so. Back in 1960, tractor buyers had the choice of gasoline, a lower powered diesel, or sometimes LPG. Allis-Chalmers changed all that with the D19. A power race was raging in the American tractor market, and A-C had no choice but to keep up. At the time, its most powerful tractor was the 53hp D17, but buyers wanted more—a lot more. It soon became clear that the projected 60hp D18 wouldn't be enough to satisfy them, so an eleventh hour project was put in hand. In theory, the existing 262ci diesel could have been enlarged to 290ci, until someone pointed out that a turbocharger would have the same effect on power, and be quicker to develop. It was an innovative move, as turbos were common on trucks at the

SPECIFICATIONS

Engine	262ci (4.1 liter)
Transmission	8-speed
Horsepower	62hp/67hp
Weight	6,835lb (3,107kg)
Numbers built	10,597
Wheels & tires	6.50x16/15.5x38
Price new	$5,834

Left: Transmission-wise, there was little new on the D19. It used the four-speed gearbox plus two-speed Power Director, to give eight forward speeds in all. A second clutch ran at 70% of engine speed, thanks to two reduction gears.

1963 ALLIS-CHALMERS D19 TURBO

D19 Turbo re-established A-C as a maker of top-power tractors.

Above: More power meant more instruments, or so it seemed.

Left: Massive 38in tires (a belated catch-up with the competition), emphasized the D19's size.

Above: Cyclops, the one-eyed giant. Twin headlights (or even any headlights) still weren't mandatory on tractors.

time, but no one had yet used one on a production tractor. The effect was dramatic. Power was boosted by 25%, to 67hp at the PTO and 62hp at the drawbar. In these days of 300hp tractors, that doesn't sound like an awful lot, but in 1961 it was enough to promote A-C up among the front runners of the power race. As if that weren't enough, the gasoline D19 had over 70hp. The D19 was actually only in production for three years, a reflection on how quickly the power race was running. By then, 100hp machines had eclipsed it, but the D19 Turbo remains a true pioneer.

Right: The D19's six-cylinder 262ci diesel engine wasn't actually new, being inherited from the D17. It could be traced back to the original 230ci unit of the WD45D, bought by A-C from Buda.

Far Right: Although it made a huge impact in 1961, within a few years the D19 had been overtaken by a new generation of 100hp tractors.

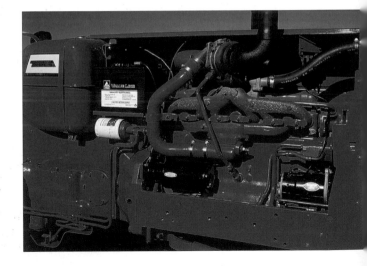

Above: No turbo, but the brand new 426ci direct injection diesel was powerful enough to act as the Allis flagship. In updated form, it powered Allis tractors for many years.

Right: As A-C's most powerful machine yet, the D21 had aggressive, squared-off styling. Launched with 103hp in 1963, it was turbocharged in 1965 to give 128hp.

ALLIS-CHALMERS   One-Seventy

Above: From 1965, Allis-Chalmers replaced its entire range of tractors with the squared-off 100-series. One Ninety replaced the D19 in that year, and this One Seventy replaced D17.

Right: The One Ninety offered a cab option, reflecting increasing concern about roll-over accidents. Power was from a Perkins 236ci diesel, or A-C's 226ci gasoline.

Left: Allis Chalmers 4W-305. The company came late to four-wheel-drive, not producing its own in-house design until 1976 with the 7580. This 1982 4W-305, powered by a 731ci 6120T turbodiesel, was built only a few years before A-C closed as an independent manufacturer.

Right: This 7020 was the base model of the original 7000 range, but in 1975 was supplanted by the 106hp 7000, which replaced the old 200. Both of them used the Acousta Cab, which Allis claimed was the quietest in the business, and variations on the Power Director transmission—12 speeds on the 7000, 20 speeds for the others. In all 7000s, the hydraulic system was load-sensitive—pressure and flow automatically adjusted to the work being done.

Above: In the '70s, Allis gave up on making its own small tractors and imported them instead—from France, Romania, and Italy. This 6140 is really a Toyosha from Japan.

Right: Fruit of the AGCO revival, the 8630 still used an air-cooled Deutz diesel, of 120hp. Options included front wheel assist (a part-time four-wheel-drive) and 36-speed transmission.

Case

Left: Case has been the great survivor—from 19th century maker of threshers to modern multi-national.

Left: 9-18 was the smallest Case Crossmotor, so called because of the transversely mounted engine. 10-18 shown here was very similar, but with slightly more power.

Above: Fully enclosed engine for this 12-25, as well as distinctive twin fuel tanks. It was built between 1914 and 1918, Case's first attempt at a smaller tractor.

Classic Profile

CASE

If two horizontal cylinders were the classic John Deere layout, then this is the Case equivalent—the Crossmotor. It got the name from its four-cylinder engine, mounted tranversely across the chassis. Case made several different Crossmotors, but the 15-27 was by far the most successful. Rugged and dependable, it helped usher in the new age of truly practical tractors that would start easily and work all day, year on year, without constant attention. Look at the dimensions—the axles

SPECIFICATIONS

Engine	382ci (6.0 liter)
Transmission	2-speed
Horsepower	27hp/15hp
Weight	6,350lb (2,886kg)
Numbers built	17,628
Wheels & tires	6.0x32/14.0x32
Price new	$1,682

Above: That's the attraction of old tractors: almost every part is open to view, its function immediately obvious.

Right: The distinctive curved arm served to support the clutch assembly, which was housed inside the belt pulley. Also visible is the four-cylinder distributor.

Crossmotor. The tranverse mounting of the 15-27's engine is obvious in this picture.

1919

Above: Over 17,000 sold in five years—this was the most successful Case Crossmotor.

were 2.75in in diameter, and the three-bearing crank 2.5in across—the machine was built to last. Unusually, the engine was a modified car unit, inherited when Case bought the Pierce Motor Company; that accounted for its advanced specification of pressure oiling and overhead valves. It was redesigned for tractor use though, the cylinders and upper half of the crankcase being cast as one. To ease servicing, the cylinder head was removable, and ports in the crankcase allowed access for cleaning of the coolant passages.

You could buy four Fordsons for the price of a 15-27, yet Case sold over 17,000 of them—a testament to its qualities.

Right: Oil sight glass. Everything about the 15-27 was aimed at ease of use—JI Case & Co realized that this was a key way to sell tractors.

Far right: Even the controls were designed to be operated from the driver's seat—something not all early tractors could boast.

Left: 15-27 wasn't the biggest Crossmotor by any means. The 22,000lb 40-72 used a 1,232ci engine that consumed 50 gallons of fuel a day.

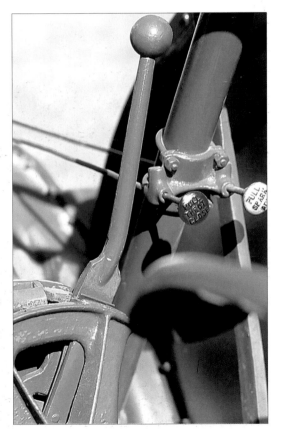

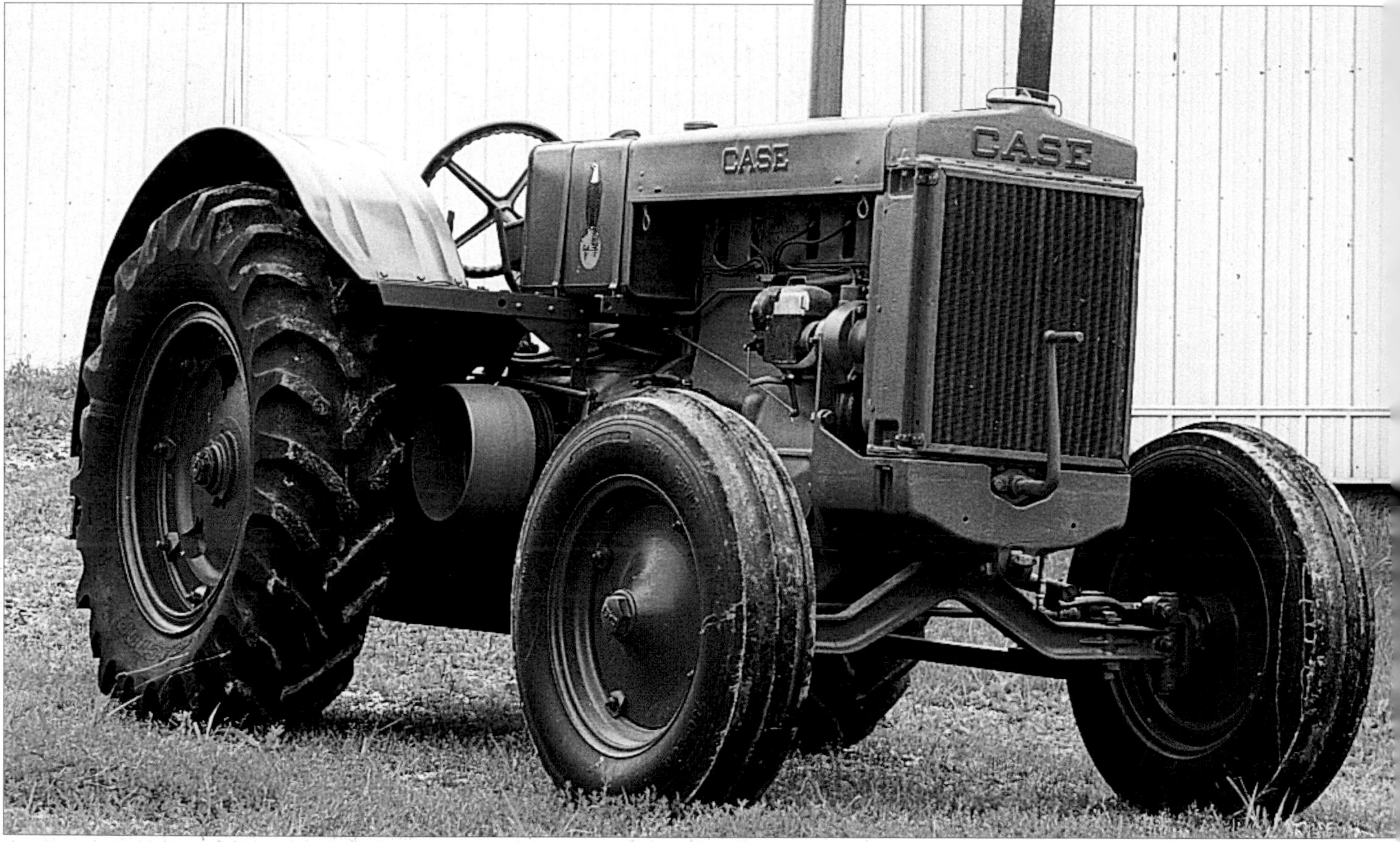
CASE
CASE

Left: Case Model L of 1929, which spelt the end of the Crossmotor era and a return to a more conventional design. The L was quite advanced—much lighter than the old Crossmotor, thanks in part to a unit-construction frame, with a three-speed gearbox. A power take-off was another new feature, which John Deere, for example, didn't have. Unusually, final drive was by chain between the axle and forward-mounted differential. A long-lived tractor, and as the updated LA, it stayed in production right up to 1953.

Right: Case didn't just make small and mid-size Crossmotors. This 25/45 left the factory in 1925, and nor was it the biggest. The 40/72 was certified by Nebraskan tests as producing almost 50hp at the drawbar, and over 90hp at the belt. The 25/45 was updated as the T model in 1928, by which time nearly 50,000 Crossmotors of all types had been sold.

Left: Case RC was easily recognizable by its distinctive "chicken roost" steering arm. It used a Waukesha four-cylinder engine giving 17hp at the belt, and a three-speed transmission.

Above: Case Model C, a scaled down L, with a 324ci four-cylinder engine. It was aimed directly at the Fordson, which was much cheaper but less stable and used more fuel.

Above: The RC, Case's answer to the successful Farmall in the sub-20hp market. The characteristic "chicken roost" steering arm marks this general purpose machine.

Right: It looks very different, but the R was no more than a standard track version of the RC, with identical engine and transmission. This one has the post-1938 "sunburst" grille.

Below: In 1939, Case replaced its entire line-up with the new Flambeau Red machines in place of the familiar Case grey. This is the S Model, a two-plough machine.

Right: Case sold around 300 Model Cs a month, while Ford was selling 2,000-odd Fordsons. But the figures are immaterial—Case survived where many rivals did not.

Left: A two-plough tractor, the Case Model S replaced the light weight R and RC. It was all-new, with its own 165ci four-cylinder engine giving 32hp at the PTO, 28hp at the drawbar. Speeds in the four-ratio transmission ranged from 2.5mph to 10.3mph according to Nebraska tests, and the S weighed around 5,000lb. This is the S standard tread, but there was also the SC row-crop, SO orchard model and SI industrial.

Right: Another of the 1939 new generation, the three-plough D replaced the Model C. The standard power unit was a 259ci gasoline engine, but for the first time there was an LPG (liquid petroleum gas) option. A PTO was standard, as was the mechanical Motor Lift, later replaced by full hydraulics. Production ended in 1953.

Left: Baby of the family, the little single-plough V Model was the Case answer to the Allis-Chalmers B and John Deere L. Heavier and more powerful than either, it sold well.

Above: This was the first tractor to use Eagle Hitch, the Case answer to the Ford-Ferguson three-point hitch. No draft control, but it did give snap-on attachment of implements.

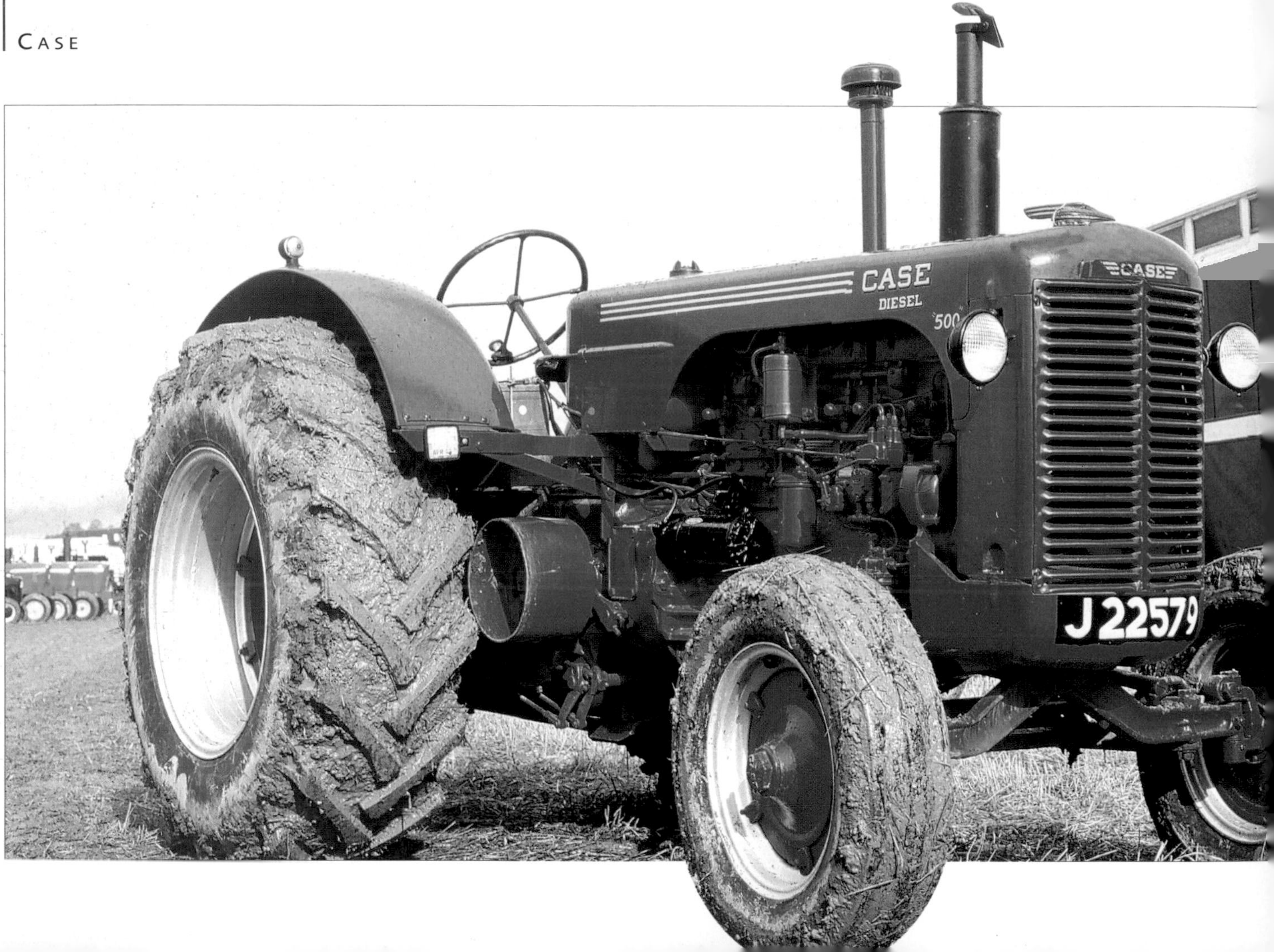
CASE
DIESEL
500
CASE
J 22579

Left: A milestone for Case tractors—their first diesel. In fact, the new 500 of 1953 was little more than the LA it superseded, with a higher rear axle ratio to suit the higher running speed of the diesel. There were still gasoline and LPG options though, which some farmers preferred. It was also the first Case with a number, being followed by the mid-range 400, big 600 and smaller 300. But unfortunately none sold as well as Case hoped.

Right: All-new for the 500, its 377ci six-cylinder diesel produced a respectable 64 belt hp at Nebraska.It was a sturdy engine, with a main bearing between each connecting-rod, and indirect injection Lanova-type combustion chambers.

Above: Case 400 Diesel in smooth sided Orchard bodywork. The all-enveloping shape was designed to prevent snagging on branches and damage to valuable trees.

Right: Case entered the diesel age with the 1953 500. It featured power steering as well as the new diesel but was the last Case finished in Flambeau Red only.

Left: Comfort King was a step forward in cab design with the 930 of 1962. The driver's platform was rubber mounted, to set new standards of comfort.

Above: Catch-up time. New squared-off styling denotes the 1957 900, which could trace roots back to the original L. The smaller Case 200, 300, and 400 were genuinely new, however.

Classic Profile

In the 1960s horsepower race, there were two ways to keep up—turbocharge, or enlarge. Allis-Chalmers had pioneered the former in 1961, but Case built one of the industry's first 100hp two-wheel-drive tractors via the simple method of more cubes. Few would have predicted that in the early postwar years, when the Case tractor line looked sadly outdated. Apart from the little VA, every model had roots in the 1920s. None of these tractors offered a diesel engine, or live power take-off, draft control or power steering. But from 1953, the company played catch-up, with an intensive program that saw it back in the forefront of tractor technology by the end of the decade. It was also heavily in debt, but that's another story.

First came a new diesel engine in four or six-cylinder form, fitted to the 500, which was an updated L from 1929. But the 400 that followed was genuinely new, with power

Above: More cubic inches equaled more power—a tried and tested approach that worked for Case.

Above: "Outta my way!" Aggressive, up to the minute styling marked out the new generation Case tractors.

1966 CASE 1030 "COMFORT KING"

SPECIFICATIONS

Engine	451ci. (7.0 liter)
Bore x stroke	4.375 x 5.0in
Transmission	8-speed
Speeds	2.0-16.2mph
Horsepower	88/102hp
Weight	9,335lb (4,200kg)
Fuel	13.24hp/hr per gall

Right: Who'd have thought it? In the early '50s, Case had fallen far behind its rivals—a dozen years later, it had the most powerful two-wheel-drive tractor on the market, with features to match. Sadly, it had overreached itself, and takeover was the end result.

Classic Profile

steering, an eight-speed transmission and the 49hp diesel option. It also had Eagle Hitch (Case's own three-point hitch, but with no draft control), and chain final drive was finally ditched in favor of shafts and gears. By 1960, the whole range apart from the big 70hp 910 had been substantially updated. The 30 series appeared that year, and in 1962 the rubber mounted Comfort King cab was launched.

So far so good, but Case couldn't afford to stand still, with power outputs increasing year on year. Its answer was this, the 1030. Instead of turbocharging the six-cylinder diesel, it increased the bore by a quarter-inch, to give a capacity of 451ci, while the five-inch stroke was unchanged. The result was 102hp at a rated 2,000rpm, with 88hp at the drawbar. This was the most powerful Case yet, and also one of the most powerful two-wheel-drive tractors on the market. Case had finally left its 1920s roots far behind.

Below: "30" denoted the 30 series, which was unveiled in 1960 as the 630, 730, and 930— the 1030 shown here came later. Engines were largely unchanged, but transmissions were updated to include triple ranges, shuttle-shift, and both hand and foot clutches. By spending a lot of money on new tractors in the '50s, Case was almost penniless by the end of the decade—it was saved by takeover.

Right: Welcome to the power generation. The 1030 was the first 100hp Case. Rather than turbocharge its diesel as Allis-Chalmers had done, Case chose to increase capacity to 451ci. That extra quarter-inch on the bore produced 102 PTO hp at 2,000rpm.

Above: Case jumped into the big four-wheel-drive market with the 1200 Traction King. This 2670 was the 1972 version, with 221 PTO hp from its intercooled 451ci diesel.

Right: 1969, and the 70-series Agri-King replace the 30-series Comfort King. All had rubber-mounted platforms and hydrostatic power steering. But what was it about "kings"?

Left: Patriotic bi-cenntenial paint job for the 1976 1570, the "Spirit of '76" special. It was the largest, most powerful two-wheel-drive tractor you could buy—180bhp at 2,100rpm.

Above: A 90-series superseded the 70-series in 1978, the bigger ones still using the in-house 504ci diesel, though micro-electronics appeared for the first time.

Left: Case Magnum 7130, 1988, part of the rationalization following the merger with International in 1984. All of International's range of 95hp+ tractors were dropped, in favor of the Magnum and Case's own 94-series. This 7130 has a 264hp 532ci diesel, and 24-speed transmission.

Below: Front-wheel-assist for this Case-International (as they were now known) from the same era. Once the huge range of machines had been rationalized, Case-IH found itself in a strong position to emerge from the troubled 1980s, with a wide spread of dealers and freshened-up range of tractors.

Classic Profile

Case was an early player in the four-wheel-drive market. As tractor power approached and surpassed 100hp, it became increasingly obvious that two-wheel-drive wouldn't always be enough, however chunky the tires and clever the transmission. Four-wheel-drive was the only answer, and Case was right there with the 1200 Traction King. It was the company's most powerful machine yet, using a turbocharged version of the 451ci diesel that had first seen service in the 1030. In this guise, it produced 120hp—plenty for this four-wheel-drive, four-wheel-steered 16,500lb machine.

In 1969, the Traction King was updated as part of the 70-series range, with clever steering (hydrostatic for the front wheels, hydraulic for the rear pair), that allowed a choice of crab steering, combined front and rear, and front or rear alone. Through the '70s, further updates brought yet more power from the Case diesel, now boosted to 504ci and with an intercooler—176hp for the 1972 2470, 221hp for the 2670. But that wasn't enough for a market in which giant four-wheel-drive tractors were becoming more mainstream year by year. The company's 504ci had reached its limit, so for the 300hp

SPECIFICATIONS
(9380 QUADTRAC, 2002)

Engine	855ci (14.0 liter)
Bore x stroke	5.6 x 6.1in
Horsepower	360hp
Transmission	12-speed
Weight	43,750lb (19,688kg)

Left: The Steiger brothers had pioneered a new breed of tractor, with four-wheel-drive and an articulated chassis. Neither idea was new on its own, but together, and combined with a big turbodiesel, it spawned a whole new generation of super-tractors. This has become the standard layout for 200hp+ machines.

Opposite: First manufacturers began fitting double wheels in an effort to improve traction, then triples, as on this 9280. For the future, more developments on the rubber crawler theme, pioneered by Caterpillar, provided an alternative.

Right: Case bought Steiger in 1986, which brought in a well respected maker of big four-wheel-drive supertractors. At first, the Steiger badge was dropped, but was later reinstated.

Left: Look closely: there's no sign of the Steiger name on this "Case International 9280." But Case-IH soon realized what an asset the Steiger name was, and reinstated it for the 9300 series, which replaced the 9200 in the 1990s.

Right: No pretense at styling finesse in this '80s four-wheel-driver—or should that be 12-wheel-driver?

2870 Traction King, they turned to Scania for a suitable diesel engine.

So by the mid-1980s, Case had been long-established in the four-wheel-drive market. That didn't stop them buying the Steiger concern in 1986. This gave access to more four-wheel-drive knowhow, not to mention the famous Steiger name, which was (and is) highly respected among American farmers. Case-IH (as it now was) continued with the established Steiger line, albeit in its own colors. Oddly, for the first few years after the takeover, Case seemed determined to drop this famous name altogether—look closely, and it doesn't appear on this 9280. But after a few years, Case seemed to realize what an asset that name was, and revived it for the 9300 series in the 1990s.

Right: More power was all very well, but turning it into useful traction was something else. This Case 9280 used triple wheels.

Below: Now Case International (after the 1986 takeover), and the Magnum 7000 series was the biggest conventional tractor in the range, half a class down from the super-tractors.

Right: By 1997, there was a choice of five Magnums, all using the Case 504ci six-cylinder turbo diesel, from 155hp to an intercooled 264hp. All had four-wheel-drive.

3

Ferguson
Massey-Ferguson

Left: Massey-Ferguson, the marriage of two famous
names to make another great survivor.

Classic Profile

Harry Ferguson's three-point hitch revolutionized tractor design, and by the early 1930s he had perfected it. Just one problem: Ferguson was a freelance operator, so he had to find a tractor manufacturer willing to take the hitch on. It was perhaps typical of the man that instead of hawking his new idea around the factories of the world, he designed a new tractor to suit it. David Brown, an English engineering firm, though with no tractor experience, agreed to build it, and the "Ferguson-Brown," the first tractor to bear the Ferguson name, started rolling off the production lines in 1936.

If ever a tractor looked revolutionary, this was it. The Ferguson-Brown had a wide track, low seat, and streamlined fuel tank that set it apart from taller conventional tractors. But apart from the three-point hitch it was really quite conventional, with a small four-cylinder engine bought in from Coventry Climax—the prototype used a Hercules unit, and after only 250 tractors had been made, David Brown took over making the engine itself, taking the opportunity to increase capacity to 133ci. But the three-point hitch alone was sufficient to accord the Ferguson-Brown landmark status. It gave quick and easy hitching; automatic draft control; and it prevented the tractor from tipping over backward, which could often prove fatal to the driver in those days before safety cabs. Sadly, the Ferguson-Brown joint venture lasted only two years—Harry went in search of a new business partner.

SPECIFICATIONS

Engine	133ci (2.1 liter)
Bore x stroke	3.25 x 4.0in
Transmission	3-speed
Speeds	1.6-4.9mph
Horsepower	20hp
Wheelbase	69in (1,725mm)

Right: As well as the three-point hitch, the Ferguson-Brown had adjustable wheel treads, and pneumatic rubber tires were an option. There was also a shorter wheelbase orchard version, with the front axle mounted further back, and an industrial model with pressed steel wheels. The Ferguson-Brown was also known as the Ferguson Model A.

Left: With its low stance and streamlined looks, the Ferguson-Brown was a very distinctive looking tractor for its time. David Brown was the first of Harry Ferguson's business partners—Henry Ford, the Standard Motor Co, and Massey-Harris would all follow.

Left: Ferguson TE20—the "Little Grey Fergie"—was one of the best known English tractors of all time, complete with three-point hitch.

Right: The Massey-Harris Pony was painted gray and sold as a Ferguson after the 1953 merger, but not for long.

Below: Although it resembled the American-made Ford 9N, the TE20 was quite different underneath. It was built by the Standard Motor Company for Harry Ferguson.

Above: Ferguson's hitch made the TE20 a very adaptable little machine. Thousands were built, and it was exported to America as the TO20.

Left: There was a whole range of variations on the TE20 "Little Grey Fergie" theme. It was updated as the gasoline Standard-engined TEA20 in 1948. The TED was powered by TVO (a blend of kerosene, not taxed in Britain), TEH lamp oil, and TEF diesel. The latter power unit is shown here, with a dieselized version of the Standard gasoline four-cylinder engine.

Right: The 35 was similar to Harry Ferguson's Detroit-built TO35, which preceded it by three years. After being rebuffed by Ford, Harry determined to go it alone.

Above: From 1957, every tractor to come out of a Ferguson or Massey-Harris factory was badged "Massey-Ferguson," and painted red over gray. This is an M-F 35.

Right: American-made Ferguson 35, using a 134ci Continental engine. Gray color scheme indicates that this is a pre-rationalization machine.

Left: The M-F 50 was really an enlarged version of the American-made Ferguson TO-35. A 2-3 plough tractor, it was powered by Continental gasoline or Perkins diesel engines.

Right: Massey-Ferguson needed a big tractor for the American market, but had nothing suitable. The quick answer was to bolt M-F badges onto a Minneapolis-Moline Gvi. That's what they did with the M-F 95 Super, pictured right.

Left: New Generation. In 1964, Massey-Ferguson replaced all its smaller machines with the 100-series Red Giants. This 135 was the smallest, made in England but exported to the USA.

Above: A Perkins diesel engine was by far the most popular option for the '60s Massey-Ferguson 135. In Europe, fuel was more expensive than in the US, and diesel made more sense even for the smaller tractors.

Above: This 165 was at the top of M-F's range of smaller tractors. It had much in common with the 135 and 150, but was powered by a 50hp 203ci Perkins diesel.

Right: Although it looked very different, the M-F 150 (later the 152, shown here) was really a restyled 50, with the same Continental or Perkins engine options.

Left: A well-used Massey-Ferguson 1130. Sold between 1965 and '67 (this one's a '67), it used Perkins' familiar 354ci six-cylinder diesel, now in 120hp turbo guise. This was Massey-Ferguson's first turbo tractor, though a naturally aspirated 1100, with 90hp, was also offered. Both these tractors were built in the US, part of M-F's rationalization plan to make its various tractor factories specialize in a certain class of machine. More big tractors were sold in the States than anywhere else, hence the logic of building all of M-F's big machines there.

Right: Sitting in a crop of alfa alfa, this Massey-Ferguson 1080 is a non-turbo tractor, using an 80hp 381ci Perkin. Both these tractors were built shortly before cabs—or at least, roll-over protection—became common.

Opposite: Bigger Massey-Fergusons such as this 1135 were built in North America rather than Europe—at the time, that's where the main market and expertise for big machines lay. The 1135 was a high powered two-wheel-drive tractor, with 121hp from its turbocharged Perkins 354ci six.

Below: Mid-range tractors were often built at M-F's plant at Beauvais, in France. This 3090 was part of the 3000 series announced in 1986, which ranged from a 71hp 3050 to the 107hp 3090, all with electronic control of the three-point hitch.

Left: Massey-Ferguson's 3000-series of the 1980s and '90s was its mid-range tractor, built at the Banner Lane plant in England, which could trace its Ferguson connection back to the TE20 days. It still produces M-F tractors today. All the 3000-series were powered by Perkins diesels, initially from the 71hp 3090 to the 107hp 3090. The 3095 shown here joined the range soon after.

Right: "Autotronic" on the cab door signifies automatic operation of the differential. Front-wheel-assist was also standard on the higher spec 3000-series Massey-Fergusons.

Left: 3090 working hard, hauling a trailer-full of grain. It wasn't the top M-F tractor of the time. The 2000-series (announced in 1979) covered the 110-150hp range, the most powerful using an intercooled Perkins six-cylinder diesel. All tractors had a push-button 16-speed transmission.

Right: In 1987, M-F replaced the 2000-series with the 3600, which brought slightly more power. Lower spec 3600s soon replaced the 3000-series tractors as well. M-F was now making its own big tractors.

Below: By the early '90s, the 3600 series was state of the art, with 32-speed (forward or reverse) Dynashift transmission and Datatronic, which allowed automatic operation of the front-wheel-assist, the PTO, and differential lock.

Right: 3630, one of the lower-powered 3600-series, used the latest Perkins six-cylinder Quadram direct injection diesel. Quadram referred to a combustion chamber in the piston crown, which Perkins said evened out combustion pressure peaks and improved torque at low speeds.

Ford/Fordson

Left: Henry Ford revolutionized the tractor market with mass production. Tractors still bear his name today.

Classic Profile

With the Model T car, Henry Ford made motoring cheap —mass production was the key, and an almost fanatical attention to cost cutting. With the Fordson F, he did exactly the same job for tractors.

The Fordson—he couldn't use his own name on a tractor, because someone else with the same name had got there first—was light, simple, and easy to understand. Above all, it was cheap. When most tractors cost over $2,000, the Model F was announced at $795. And in true Model T

SPECIFICATIONS

Engine	251ci (3.9 liter)
Transmission	3-speed
Horsepower	12hp/22hp
Weight	2,710lb (1,232kg)
Numbers built	674,476
Wheels & tires	28x5/42x12
Price new	$230-795

fashion, as more were made, and costs came down, Ford passed his savings on to the customer—at one point, you could buy a brand new Fordson F for just $230, a staggeringly low price.

Henry had been brought up on a farm, and had a good idea of what small farmers needed, even if they didn't themselves until it was put in front of them. He had actually been experimenting with tractors for ten years before the F went on sale, using Model T parts.

Left: One drawback of the little Fordson was its short wheelbase—it was prone to rear up and tip over backward, which could be fatal for the driver.

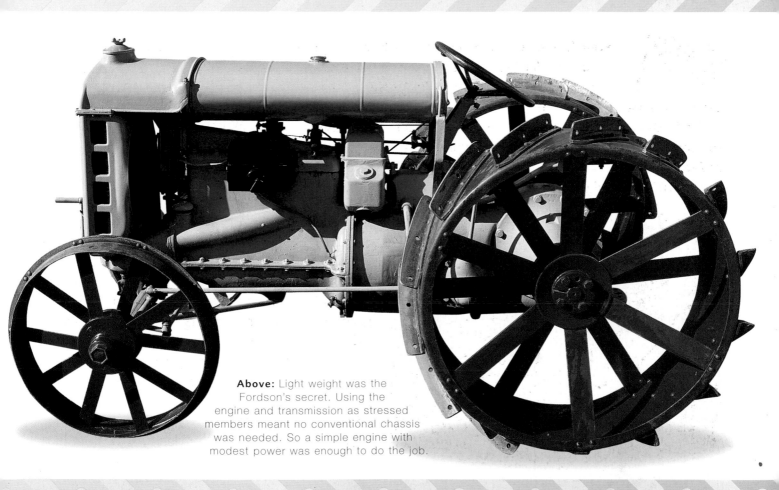

Above: Light weight was the Fordson's secret. Using the engine and transmission as stressed members meant no conventional chassis was needed. So a simple engine with modest power was enough to do the job.

"Fordson" didn't become
"Ford" until 1939.

The F made good use of that experience, ending up as lighter and easier to handle than most rivals, which were big plus points for farmers who were contemplating their first ever tractor purchase. And they did so in droves—over half a million Fordson Fs were sold in 12 years. Oddly, Henry Ford stopped production in 1929, clearing the decks for the new Model A car, but the Fordson lived on, produced in England and Ireland. A legend had been born.

Left: Without doubt, the Fordson F was one of the most significant tractors of all time. Not technically, for (unit construction apart) it was by no means advanced. But what it did do was put mechanized farming within the reach of thousands of small family-run farms—it opened up a whole new market for tractors.

Right: Milestone though it was, the Fordson F wasn't without its faults. Reflecting its price, the F was a little crude. That hardly mattered for its first decade, but by the late 1920s, farmers were expecting more adaptability. The Farmall had shown the way forward, though Ford chose not to update the F.

Left: Fordson F, 1929, and essentially the same as the 1917 original. Ultimately, that was the problem. Henry Ford was more interested in cutting costs than updating the F.

Above: Outdated in some ways by 1929, but still a best seller. Only six years after the launch, Ford was claiming that 75% of all tractors on US farms were Fordsons.

Henry Ford's Fordson F—the tractor that revolutionized farming—had much in common with his Model T car. Not only was it cheap and simple, but Henry refused to update it, preferring to stick to his "stack 'em high, sell 'em cheap" philosophy. But in the meantime, the opposition had acquired more power and sophistication—the Fordson needed to catch up. A shift in production to Ireland provided the ideal opportunity, and six months later, in 1929, the Fordson N was announced. The changes were hardly radical

SPECIFICATIONS

Engine	267ci (4.2 liter)
Bore x stroke	4.13 x 5.0in
Transmission	3-speed
Horsepower	29hp (PTO)
Weight	3,600lb (1,636kg)

Above: In 1939, Henry Ford began using his own name on tractors, but in Britain they carried on with the Fordson badge.

Right: An increase in rated speed from 1000rpm to 1100rpm brought 23hp on kerosene, 29hp on gasoline, according to Nebraskan tests.

—a slightly larger engine and increased rated speed (from 1000 to 1100rpm) boosted power to 29hp at the belt on gasoline. Opt for kerosene, and you got 6hp less. The basic Model T ignition system was ditched in favor of a high-tension magneto, and the orchard style rear fenders were made standard. Cast front wheels were stronger and heavier than the old spoked ones and the front axle was heavier-duty as well. It was just as well that the Fordson N had more power than its predecessor—it weighed nearly half a ton more.

Classic Profile

But the N did not have an easy time of things. It was launched into a worldwide tractor recession, and the Cork production plant was expensive to run, too far from the main tractor market in England. By 1933, production had been moved again, to Ford's Dagenham factory—the N emerged with blue paint and conventional fenders. There were new options over the next few years, such as pneumatic tires, and electric lights and starting, but the Fordson N was still no more than an update of Henry Ford's 1917 original.

Right: Bright blue bodywork marks out this as one of the Dagenham built Ns. Ford's English factory took over tractor production from the Cork plant in 1933, and carried on building the now outdated N until 1945.

Left: Dagenham N with steel wheels and spade lugs. Ford's British arm was offered the new 9N tractor to build, but it refused, preferring instead to persevere with the N and its own comprehensive update, the E27N.

Right: Fordson N with Roadless conversion. These half-tracks were popular for a variety of uses—in forestry as well as farming, and for towing lifeboats. Roadless Traction Ltd converted many Fordsons, as well as other tractors and trucks.

Above: A four-ringed badge denotes Perkins six-cylinder diesel power for this E27N. It was a great improvement on the elderly Fordson gasoline power unit.

Right: Fordson E27N—"E" stood for English (it was built at Ford's Dagenham factory), "2" was the horsepower (not quite enough any more), and "N" showed the lineage.

Left: For 1952, Ford in England replaced the ageing E27N with the New Major—now fully up to date, with six-speed transmission and Ford's own diesel engine.

Above: More updates followed—the 1959 Power Major diesel enjoyed a 22% power boost and was a genuine five-plough machine. Later sold through US dealers as the 5000 Diesel.

Classic Profile

The Ford 9N was as big a milestone as the original Fordson F. Like the F, it was small, light, and cheaper than most of the competition. Unlike the F, it had a revolutionary new feature—the Ferguson three-point hitch. Through a mutual friend, Harry Ferguson demonstrated his hitch in 1938 to Henry Ford, who was so impressed that he immediately agreed to build an all-new tractor to suit it.

The 9N was the result, a joint effort of Ford and Ferguson engineers who worked so hard that the new model was on sale within a year. Ferguson's hitch was the key selling point: it allowed quick attachment of implements; it had draft control and prevented the tractor tipping over backward;

SPECIFICATIONS

Engine	120ci (1.9 liter)
Transmission	3-speed
Horsepower	13/20hp
Weight	2,340lb (1,064kg)
Numbers built	99,002
Wheels & tires	4.00x9/8.00x32
Price new	$585

Right: Nice detail work. The 9N was an exceptionally clean and modern looking tractor.

Left: By 1939, Henry Ford was able to use his own name on a tractor, though the British carried on building what they called "Fordsons."

Above: Just as the first Fordson F relied upon light weight as its secret weapon, so did the new 9N. It weighed only 2,340lb, while a typical two-plough rival tipped the scales at 3,700lb. The result was an excellent power-to-weight ratio, and the 9N was said to be capable of the same work as a tractor of twice the weight and power.

Classic Profile

Below: Ferguson's three-point hitch was a major part of the 9N's appeal and a real milestone in tractor design.

and it used some of the implement's drag as downforce on the tractor's rear wheels, improving downforce. All this ingenuity, plus Ford's production expertise, should have meant a marriage made in heaven. Unfortunately, the original handshake agreement ended in a multi-million dollar law suit.

Right: The 9N was small, but it could still handle heavy loads such as this scraper.

Below: If it looks right, it is right—the old engineering adage works as well for tractors as it does for anything else.

Revolution in gray. The 9N was just as much of a milestone as its Fordson forebear, though for different reasons.

Below: Ford 8N, a modernized 9N which marked the break with Harry Ferguson in 1947. Henry Ford II had discovered that Ford was losing money on every 9N it made, thanks to the marketing agreement with Ferguson. Now with direct control of distribution, Ford found tractors profitable.

Right: A less than pristine 8N, though perhaps in more "authentic" condition than the one below! There were 20 improvements over the 9N, including a four-speed gearbox, better brakes, and a position control for the hydraulics. It was a huge success—over 100,000 were sold in the first full year.

Below: By the early 1950s, the Ford 9N/8N design was nearly 15 years old. Buyers wanted more power and sophisticated transmissions. They got it with the NAA of 1953, complete with overhead valve engine and new bodywork. Still no diesel option though, reflecting that this was a US-built tractor.

Left: NAA buyers did have a modern 134ci gasoline engine, which carried over to the 1954 600-series. An 800 was announced at the same time, with a 172ci version of the same engine.

Left: Ford's 600 series replaced the NAA after only a year. The 640 came with a basic four-speed transmission; the 650 was a five-speed, and the 660 was fitted with a live power take-off as well. Still gasoline only though, so Ford imported the English Fordson equivalent to fill that gap—in a few years the British and American arms of Ford tractors would be unified. Both the 600 and the more powerful 800 later came in row-crop guise, too.

Right: Ford 6000, 1961, first fruit of the new rationalized line of US- and UK-built tractors. The 6000 was a mid-range machine, powered by a 223ci six-cylinder gasoline engine or 242ci diesel—the gas option gave 63hp at the drawbar, nearly 67 at the PTO. Ford's Select-o-Matic transmission also featured, allowing the drive to shift through all ten forward ratios without stopping. Unfortunately, it was less than durable, and was subject to several recalls and updates. This 6000 is a row-crop, a tractor layout that would be gone within a decade.

Left: New decade, new look. Ford's newly rationalized range included the three-cylinder 3000 in gasoline or diesel form, with modern squared-off styling to match the bigger 4000 (really a restyled version of the old 800), and smaller 2000, which also offered three-cylinder diesel or gas.

Below: In 1968 Ford produced its first 100hp+ tractor, the 8000. It used a 401ci six-cylinder diesel, and came with an 8- or 16-speed transmission. Ford was still trying to shake off the legacy of its unreliable Select-O-Matic system, which had proved something of a disaster in the earlier 6000.

Below: Remember the big, beefy, straightforward 8000, with which Ford sought to exorcise its Select-O-Matic demons? Ten years on, they used the 8000's 401ci diesel as a basis for the new 7810, though in detuned 86hp form. A 16-speed transmission was optional, as was front-wheel-assist (a sort of halfway house to full four-wheel-drive). Big, glassy cabs with roll-over protection were standard fare now.

Above: After concentrating on small tractors for many years, Ford finally got into the horsepower race in the late '60s. This 9000 was a turbocharged version of the big 8000, produced 131hp, and needed those twin wheels.

Right: By 1980, the 86hp 7810 was little more than a mid-range tractor, while Ford's line-up ranged from 11hp mini-tractors to 300hp+ giants. Times had changed since the Model F.

Left: While power outputs and engine sizes headed skyward, Ford didn't neglect the utility end of the market. This 70hp 5610 was typical, available in two- or four-wheel-drive, with or without a cab. It started out as the 60hp 5600 in 1976, getting a power boost in 1981 and a 16-speed transmission option two years later, when it also acquired the "10" suffix, alongside the 6610 and 7710.

Right: A 1990 6510, powered by a four-cylinder diesel of 256ci and 62hp at the PTO. It's interesting to note the use of four-wheel-drive, which by the late 20th century was trickling down to the lower powered utility tractors.

Right: Ford broke away from its four-digit numbering with the TW-series in 1979. These were two-wheel-drive flat platform tractors, with the driver sitting on top of (not astride) the transmission. Revised in 1983, they still used the 401ci diesel, in normally aspirated, turbo, or intercooled form. The TW-15 here was the mid-range 121hp machine, while the intercooled TW-35 with 170hp was Ford's most powerful two-wheel-drive tractor yet.

Above: Ford or New Holland? By the early '90s, they were wearing both badges. Ford had taken over implement maker New Holland in 1987. This 7840 was top seller of the popular mid-range 40-series.

Right: Purchase of the Versatile company brought super-tractors into the Ford-New Holland line up. Ford's own engines were used at first, later supplanted by Cummins power plants.

Hart-Parr/Oliver

Left: Oliver's six-cylinder tractors set new standards of smoothness in the 1930s and '40s.

Below: This 18-36 was Hart-Parr's mid-range tractor, the final incarnation of the original two-cylinder 15-30 which had been launched in 1918. A twin-cylinder throttle governed machine, it was joined in 1920 by the smaller 10-20 (also a twin) and three years later by the big 22-40—the latter was powered by two 10-20 engines side by side! A powered-up 28-50 appeared in 1928. By then however, the basic design was a decade old, and out of date.

Above: College friends Charles Hart and Charles Parr were tractor pioneers, building prototypes in 1902 and in production by the following year.

Left: Hart-Parr pioneered small tractors as well, though the two-stroke "Little Devil" was notoriously unreliable. But this, the more conventional 12-24 of 1918, was a success—it could do the work of a machine twice as heavy, and was soon re-rated to 15-30 horsepower, reflecting its true output.

Classic Profile

"Old Reliable" was its nickname, and deservedly so. In 1903, when most tractors were frail curiosities of limited life span, along came Charles Hart and Charles Parr with a truly reliable, practical tractor, the 22-45, later uprated as the 30-60 shown here. Other tractors broke down or wore out, or refused to start. The Hart-Parr started on demand. Other tractors could last less than a year in the field—at least one Hart-Parr 30-60 was still working 20 years after leaving the Iowa factory.

Looking more like a steam traction engine than the tractor as we know it (just like its contemporary rival, the Advance Rumely Oil Pull), the big 30-60 was powered by a massive twin-cylinder engine with a 15in bore. Each piston measured 10in across! It turned over at a stately 300rpm, pushing the complete machine along at a slow walking pace in its single gear. Not surprisingly, it also weighed over 10 tons.

SPECIFICATIONS	
Engine	2356ci (37.0 liter)
Transmission	1-speed
Horsepower	30/60hp
Weight	17,820lb (8,100kg)
Numbers built	3,445
Wheels & tires	c66in
Price new	$2,600

Left: Clearly visible here is the gear-drive for the belt pulley (exposed like many of the mechanical parts of early tractors), as well as the engaging mechanism. Big machines like this Hart-Parr were used more for belt work than hauling implements.

1913 HART-PARR 30-60

Above: As if the experienced horseman didn't have enough new technology to cope with, the Hart-Parr started on gasoline, but ran on kerosene once warmed up.

Top speed, 2.3mph, but virtually unstoppable....

Classic Profile

Nor was this the biggest tractor made by Hart-Parr. They also offered the 40-80 between 1908 and 1914, which weighed 18 tons and boasted 8ft driving wheels. If that weren't enough, the 60-100, one of the most powerful tractors of its time, weighed 26 tons and had 9ft driving wheels. Behemoths like this would soon be gone though—in the 1920s, a new breed of lighter four-cylinder tractors, easier to operate and cheaper to buy, began to take over.

It was Hart-Parr incidentally, who coined the term "tractor," or at least their sales manager WH Williams. The name stuck, being far easier on the tongue than "gas traction engine." In 1924, Hart-Parr dropped the 30-60, but by then it had done its job—it had proved without question that gas tractors could be strong and reliable.

Right: No doubt about who made this one. With their big water tanks and steaming cooling towers, these early gasoline tractors could easily be mistaken for steam traction engines.

Above: Exposed steering gear looks picturesque now, but wore out quickly in the mud and dust of field work. Compared to many contemporaries though, the 30-60 was long-lived.

Right: "30" was the claimed horsepower for pulling a plough, with "60" available at the belt pulley. However, in early tractors, these figures were often approximate.

Above: The tractor that drove Hart & Parr to leave the company they founded: partner Charles Ellis wanted to produce a smaller tractor (the 12-24), but they didn't.

Right: Largest of the "small" Hart-Parrs, the 28-50 of 1928 was an uprated version of the twin-engined 22-40. But Hart-Parr couldn't survive on its own, and a merger was imminent.

Far left: In 1929, Hart-Parr merged with the Oliver Chilled Plow Company and two other implement makers. It was the perfect match— Oliver had a modern tractor on the drawing board, while Hart-Parr had the production expertise to build it. The result was the Oliver Hart Parr Model A of 1930. However, within a few years, it was just plain "Oliver," as the Hart-Parr name was dropped. The first A was rated at 18-27 horsepower, with a single front wheel and an adjustable rear track—the rear wheels slid in or out on splines to suit different crop widths. It marked the start of a new era for both Oliver and Hart-Parr.

Left: Apart from its splined rear axles, the Model A was a fairly conventional machine, though it was soon uprated into a 28/44, with a four-cylinder 443ci engine. This came in standard tread (illustrated) as well as industrial versions. A three-speed transmission allowed speeds between 2.7 and 4.3mph.

Classic Profile

Until the mid-1930s, the typical tractor engine was a low-revving slogger with two or perhaps four cylinders. The Oliver 70 changed all that. It had a high-compression, high-revving six that brought new levels of smoothness to tractor operation. It needed to run on expensive high octane fuel, but paid that back with greater efficiency. Of course, if you insisted, you could have it in low-compression distillate form, but that seems like missing the point. The engine was bought-in from

SPECIFICATIONS

Engine	201ci (3.2 liter)
Transmission	4-speed
Horsepower	16-25hp
Weight	3,340lb (1,518kg)
Numbers built	65,000
Wheels & tires	27x4.40/59.5x6.25
Price new	$915

Right: With its sleek, long hood, the Oliver 70 looked more like a car than a tractor, especially after the streamlined "Fleetline" styling (seen here) was introduced in 1937.

Left: This Waukesha engine (though Nebraska tests recorded it as Oliver's own) introduced high-compression six-cylinder power to tractor buyers. However, again according to Nebraska, the lower compression distillate version lost little in power.

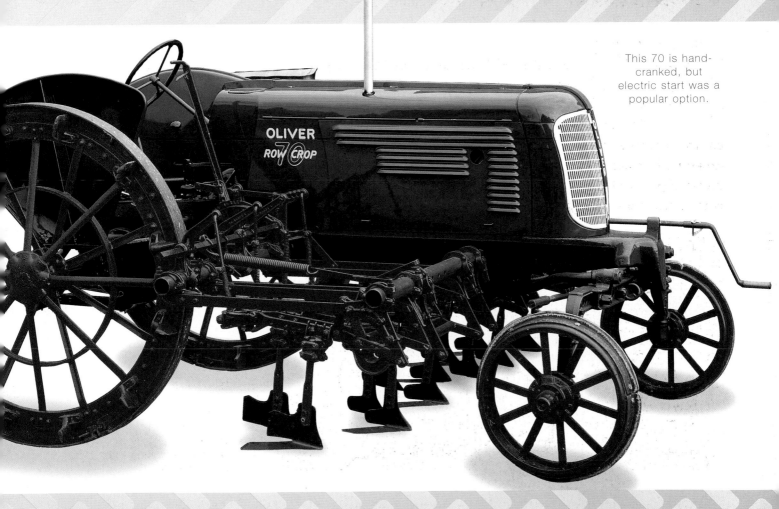

This 70 is hand-cranked, but electric start was a popular option.

Waukesha, which was to make all of Oliver's tractor engines. "Power when you want it," read one advertisement, "power when you need it—power when field conditions are right and time and help are limited. That's the power you'll find in the Oliver Row Crop 70."

Below: Like its rivals, Oliver introduced a pneumatic tire option in the late 1930s.

Above: The 70's simple dashboard comprised just water temperature and oil pressure gauges. Note six-speed transmission gate on this late-model 1943 70.

Right: Earlier pre-Fleetline styling had a boxier look, but enclosed the same smooth six-cylinder engine.

Left: Alongside the sleek, streamlined 70, the old four-cylinder Hart-Parrs soldiered on, albeit in updated form. The Oliver 90 was an upgraded version of the original Model A. Powered by the same 443ci engine, it added high-pressure lubrication, a centrifugal governor, and electric start.

Above: The 90 might look outdated, but it had a following—Oliver kept on producing it right up to 1953. In that last year, some 90s were fitted with six-cylinder engines, while production was moved from the old Hart-Parr base in Charles City to the Oliver factory in South Bend.

Below: Just as the Oliver 90 was no more than an updated Oliver-Hart-Parr, so was the 80. This time, its base was the Hart-Parr 18-27 and 18-28, which stemmed from the 1929 merger. The engine was Oliver's own 298ci four-cylinder unit. It was put through the Nebraska tests in November 1940, and produced 33 drawbar hp, and 38 at the belt. No repairs or adjustments were needed in 40 hours of testing.

Above: Oliver's engine came in high compression gasoline or low compression kerosene form. Power outputs were similar though, as the kerosene motor was slightly larger, at 334ci.

Right: Pneumatic tires for this later model standard tread 80. In 1939, a four-speed gearbox and ASAE standard hitch and PTO were made standard, with a Buda-Lanova diesel engine option the following year.

Left: Oliver's 60 was really a smaller brother to the successful 70, using a 108ci four-cylinder version of the famous six, rated at the same 1,500rpm. A four-speed transmission, pneumatic tires, and the modern engine underlined the 70-like specification. Oliver needed a small tractor to rival those from John Deere and Allis-Chalmers —the 60 fitted the bill.

Right: Introduced in 1940, the 60 was updated as the 66 seven years later. Engine size increased to 129ci, with a 145ci kerosene version, plus a diesel option. It was replaced in turn by the Super 66 in 1954, powered by Oliver's own four-cylinder engine, plus a three-point hitch, and three-speed governor. Hydra-Lectric was Oliver's version of the hydraulic three-point hitch.

Classic Profile

Was this the start of the power race? In 1947, the Oliver 88 diesel produced 43.5hp at the PTO, making it one of the most powerful tractors of its time. It was part of a new range, launched alongside the 66 and 77, with which it shared some components.

A live power take-off (PTO) and live hydraulics were up to the minute features, in what was a thoroughly modern tractor. But in its choice of power units, the 88 underlined another technical development. Since the very first tractors, gasoline or kerosene had held sway—the former gave best power, but was thirsty, while the latter was cheap but less efficient. LPG—liquid petroleum gas—was a new option, a cheaper fuel than gasoline; and the Oliver 88 was one of the first mainstream tractors to offer a diesel option—this usually

SPECIFICATIONS

Engine	231ci (3.6 liter)
Transmission	6-speed
Horsepower	29/38hp
Weight	5,110lb (2,323kg)
Numbers built	38,867
Wheels & tires	6.00x16/13.00x38
Price new	$2,810

Above: Power take-off, but still no three-point hitch on this 1948 Oliver 88.

Oliver's 88 continued the tradition of sleek, streamlined tractors begun with the 70.

Above: Oliver offered a wide range of power units for the 88, reflecting the battle for supremacy between gasoline, diesel, LPG, and kerosene.

meant less power than gasoline, but was the most fuel-efficient of all. So the 88 offered a choice of diesel or gasoline in the standard 231ci six-cylinder engine, or a 265ci kerosene unit—kerosene engines were often slightly larger, to counteract their lower efficiency. Most of these power options were available for each of the various 88 models: standard tread, orchard, industrial, high-crop, and no less than three row-crop versions. What the 88 didn't have was a hydraulic lift, but that came in 1949— Hydra-Lectric was Oliver's version of the hydraulic three-point hitch. The 88 lasted seven years, being replaced by the Super 88 in 1954.

Above: Like most of its rivals, the Oliver 88 came in a variety of forms, and the diesel version was popular with industrial users. But this wide front axle option ($78 extra) found favor with farmers in hilly country.

Right: A chunkier front grille marks out this later 88. The tractor was quickly restyled, after only 1,300 of the original streamlined 88s had left the production line.

Left: Replacing the six-cylinder 70 in 1947 (alongside the 66 and 88), the 77 offered more power and a higher rated speed. Oliver's hydraulic lift came in two years later.

Above: Ten per cent more power than the 77, and a two-range "Power Booster" transmission, giving 12 forward speeds in all. The 770 took Oliver into the '60s.

Left: It was cheaper to import smaller utility tractors than build them in the USA. Oliver's 32hp 500 and 52hp 600 were rebadged, repainted David Browns from England.

Above: 1970, and Oliver has been a White subsidiary for ten years—the end is near. This 1655 was a six-cylinder 70hp machine with an 18-speed transmission.

Below: 1655, 1970. By this time, Oliver was sharing many components with Cockshutt and Minneapolis-Moline, which had joined it under the White Motors umbrella. In a short time, White dropped all pretence of separate tractor lines, and used these once proud names as marketing tools. In the meantime, the 1655 and 1855 both used Oliver's own six-cylinder engine.

Above: Oliver still offered the GM supercharged diesel option, in this 1955. Supercharging boosted the relatively small 212ci four-cylinder unit up to 90hp at the drawbar.

Left: The 1655 came with a choice of three auxiliary transmissions, reflecting the fact that this area of tractor design was becoming at least as important as the power unit: Hydra-Power Drive, Over/Under Hydraul-Shift, or Creeper Drive. Its six-cylinder 265ci engine was Oliver's own.

Above: For its biggest and most powerful tractor—the 2255 of the early '70s—Oliver chose not to go the turbocharged route that most of its rivals followed.

Right: Instead, it bought-in the huge 473ci V8 diesel from Caterpillar. With 145hp, it was one of the last powerful two-wheel-drive tractors.

6

International

Left: Farmall was the most significant tractor
International ever made—it could do everything.

Below: International Harvester was the result of a merger of five smaller firms. To keep rival dealers happy, IH had to produce two separate lines of tractors, and this Titan 10-20 was for the Deering men. Smaller and lighter than the average tractor, it was a great success—17,000 were sold in 1918.

Right: Like many early engines, that of the Titan started on gasoline, only running on kerosene once warmed up, though water injection was an unusual feature, to prevent pre-ignition under load. A 35-gallon tank served as radiator.

Left: While Deering dealers sold the Titan, McCormick men had the 1914 Mogul 8-16. But only three years later, International replaced that with another 8-16, far more advanced, and reflecting the fast pace of tractor development of the time. It was also sold under International's own name.

Above: Small, but perfectly formed. The 8-16 was remarkably advanced for 1917, with a high-speed engine (enclosed under a hood), three-speed transmission, and a power take-off—the first commercially successful PTO.

Right: Most lightweight rivals to the International used low-revving, twin-cylinder engines. But this four-cylinder unit could spin up to 1,000rpm, a high speed for the time.

Classic Profile

The sheer impact of Henry Ford's Fordson from 1917 cannot be overstated—but that doesn't mean his rivals were doing nothing. Take the McCormick-Deering 15-30 (so named because IH's McCormick and Deering dealers still couldn't agree on which name to use). The International 8-16 which preceded it was quite advanced in some ways, but it still had a heavy riveted chassis and exposed chain drive; the chain in particular was bad news in arduous conditions, needing frequent maintenance. But in 1921, the new 15-30 answered both those criticisms, with a welded chassis and gear final drive, properly enclosed in an oil bath.

In fact, compared to the previous Titan 15-30, launched only two years previously, this one was clearly a great leap forward. Both tractors could do the same amount of work, but the McCormick-Deering was nearly 3,000lb lighter, it was two foot shorter, and easier to operate. As if

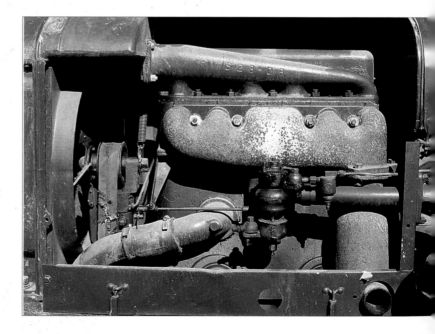

Right: A simple, straightforward two-bearing four-cylinder engine served the McCormick-Deering well. It was powerful enough for big drawbar and belt work, yet very reliable.

A smaller 10-20 version of the 15-30 was also offered, and was even more successful—over 200,000 were sold.

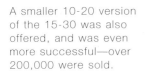

SPECIFICATIONS

Engine	382ci (6.0 liter)
Transmission	3-speed
Horsepower	15/30hp
Weight	5,730b (2,614kg)
Numbers built	128,125
Wheels & tires	34x6/50x12
Price new	$1,250

By 1921, the modern tractor was taking shape.

Classic Profile

Below: Steel lugs like this were a simple means of improving traction for steel wheels. The alternative was wheel extensions—the equivalent of twin- and triple-wheel pneumatic tires on modern tractors.

Left: Full enclosure for engine, transmission and driveline, though on this 15-30 the owner has cut a hole for quick access to the carburetor.

Right: One interesting point about kerosene fuel: although it was cheaper, it could seep past the piston rings and dilute the lubricating oil. That's why the oil had to be changed three times as often in a kerosene-fueled motor.

that weren't enough, at $1,250 it cost little more than half as much as the Titan. It was also a far simpler tractor, needing a lot less maintenance—at last, the farm tractor was turning into the reliable workhorse it had to be. And remember, the Titan 15-30 had gone on sale only two years before—that's how fast tractor technology was developing in the early 1920s.

A Fordson was still much cheaper, but that was in a different class. The McCormick-Deering was a four-plough machine, able to do useful work on medium-sized as well as small farms, with an impressive 30hp of belt power available from its 382ci four-cylinder engine. It was compact, rugged, and reliable, so it was hardly surprising that International Harvester sold over 128,000 of them in eight years.

Left: Alongside the rugged 15-30, International built this 10-20, also badged McCormick-Deering, and really a smaller version of the same thing. The 284ci four-cylinder engine drove through a three-speed transmission.

Above: The 10-20 was even more successful than its big brother, with 216,000 sold before production finally ceased in 1942. This tractor was the most popular of the pre-Farmall Internationals.

Rarely was a name so apt—the Farmall tractor really could do it all. Surprising as it might seem, this was a new idea in 1924. Until then, tractors were either small and light for cultivation, or much heavier and more powerful for belt work—driving threshers and the like—and big drawbar loads. Until the Farmall, no one had combined useful belt power (over 20hp) with the compact size and agility of a smaller tractor, which could tip toe through the fields without causing damage to valuable crops. As its name suggested, the Farmall gave the best of both worlds.

It was a long time coming though —IH had been working on the project for a full seven years before it appeared, but few would deny

SPECIFICATIONS

Engine	221ci (3.5 liter)
Transmission	4-speed
Horsepower	13/27hp
Weight	3,650lb (1,659kg)
Numbers built	135,000
Wheels & tires	25x4/40x6
Price new	$950

Left: Exposed steering gear seems like an oddly retrograde step, though it didn't affect the Farmall's success.

Right: Tricycle type twin front wheels became the standard format for row-crop tractors until the late 1960s.

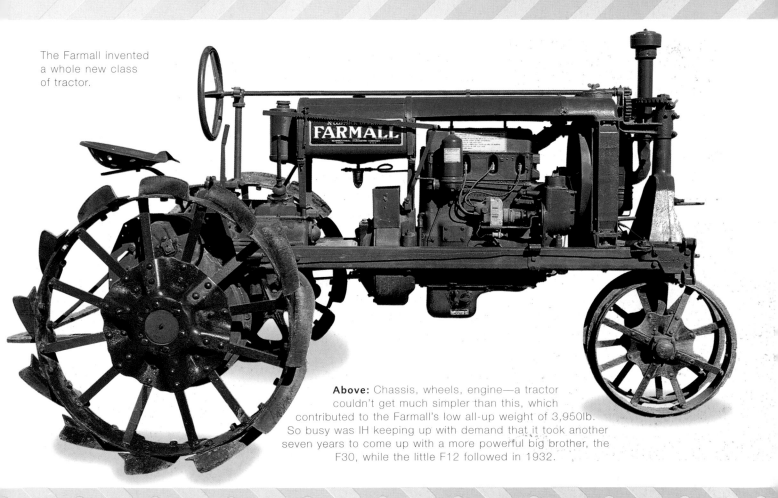

The Farmall invented a whole new class of tractor.

Above: Chassis, wheels, engine—a tractor couldn't get much simpler than this, which contributed to the Farmall's low all-up weight of 3,950lb. So busy was IH keeping up with demand that it took another seven years to come up with a more powerful big brother, the F30, while the little F12 followed in 1932.

Classic Profile

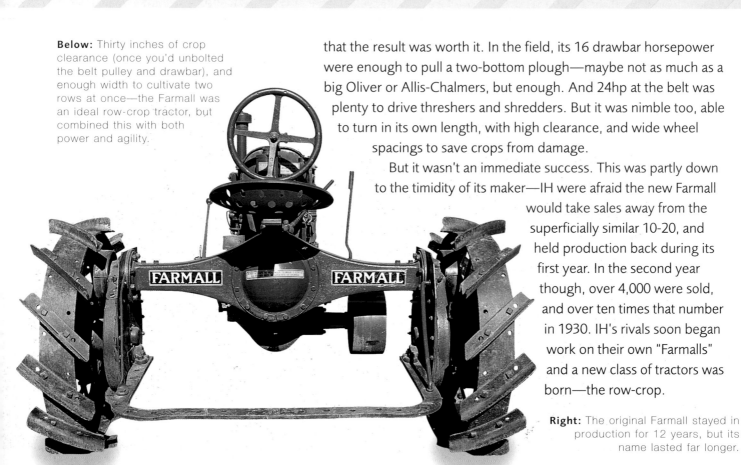

Below: Thirty inches of crop clearance (once you'd unbolted the belt pulley and drawbar), and enough width to cultivate two rows at once—the Farmall was an ideal row-crop tractor, but combined this with both power and agility.

that the result was worth it. In the field, its 16 drawbar horsepower were enough to pull a two-bottom plough—maybe not as much as a big Oliver or Allis-Chalmers, but enough. And 24hp at the belt was plenty to drive threshers and shredders. But it was nimble too, able to turn in its own length, with high clearance, and wide wheel spacings to save crops from damage.

But it wasn't an immediate success. This was partly down to the timidity of its maker—IH were afraid the new Farmall would take sales away from the superficially similar 10-20, and held production back during its first year. In the second year though, over 4,000 were sold, and over ten times that number in 1930. IH's rivals soon began work on their own "Farmalls" and a new class of tractors was born—the row-crop.

Right: The original Farmall stayed in production for 12 years, but its name lasted far longer.

Above: Many farmers wanted a smaller tractor than the original Farmall, and they got it with the F12. Its small 116ci four-cylinder engine produced 10hp for hauling, 16 at the belt.

Right: Big brother to the original, the F30 Farmall arrived in 1931. With a bigger-bore 284ci power unit, it gave 30hp at the belt, though it was less popular.

Left: Despite the success of the Farmall, some farmers still needed standard-tread tractors, and this W12 was an F12 to suit them, seen here with the pneumatic tire option.

Above: This McCormick-Deering 15-30 has extended rear wheels to improve traction, as well as spade lugs. Tractors of this age might never see tarmac roads.

Right: International built standard-tread versions of all the Farmalls, and the W-series was launched in 1940, with the same Lowey styling. W12 was a four-wheeler Farmall A, W4 (seen here), was based on the H, and W6 on the big M. But there was also a new range-topper, the 45hp W9 with a 335ci four-cylinder engine—that also came as the diesel-powered WD9.

Above: Farmall AV—"A" denoted the smallest one-plough Farmall, introduced in 1939, and "V" stood for high clearance. Even the little A got the Raymond Lowey styling treatment.

Right: The A's bigger brothers also came in high clearance V form, here with the tricycle single-wheel front end. Note the vestigial fenders, upholstered seat, and enclosed steering rod of this Lowey-styled machine.

Above: There were innumerable variations on the Farmall theme. This BMD is a British-built diesel-powered derivative of the B, which was a two-row version of the Farmall A.

Right: W6 was the standard-tread version of the big Farmall M, which itself had replaced the F30 in 1939. It was a 33hp three-plough machine—International's largest at the time.

Classic Profile

Say what you like about the early Farmalls and McCormick-Deerings, but in no way were they stylish. Compare an M-D 15-30 or original Farmall with an Oliver 70—they were rugged, boxy workhorses which made no pretence at being absolutely anything else.

This all changed in the 1940s, when IH took on renowned industrial designer Raymond Lowey to give its tractors a makeover. He started with the company logo, creating the famous red and black "IH" symbol that remained in place for many years. As for the tractors, Lowey gave them a rounded, streamlined front with integral grille. This blended into a long, clean looking hood that tapered into a graceful curve in front of the operator's station. The Farmall H on these pages is a perfect example of that clean, flowing style—tractors didn't have to be workaday collections of components just bolted together, they could look good as well.

Left: In case the driver forgot, International cast the gear pattern into the transmission. Five speeds for this 1947 Farmall H, though it wasn't long before the new breed of two- and three-speed auxiliary gearboxes would make 10-, 12- and 16-speed transmissions a reality.

1947 FARMALL H

Neat and clean—
Internationals had
never looked this
good before.

SPECIFICATIONS

Engine	152ci (2.4 liter)
Transmission	5-speed
Horsepower	20/24hp
Weight	5,375lb (2,443kg)
Numbers built	390,317
Wheels & tires	5.50x16/10.00x36
Price new	$962

Above: Generally
situated far from their
nearest dealer, farmers
were their own
mechanics.

Above: So successful was
the Farmall brand, that IH went on
using it right into the 1970s.

Classic Profile

Below: A neat styling feature, recessing the oil pressure and water temperature gauges into the hood like that. In theory, they were more in tune with the driver's sight line. In practice, these two-inch gauges were four feet away—let's hope the driver wasn't short sighted!

There were changes under the skin, too. The original Farmall had been overtaken by younger rivals, so IH replaced it with the H in 1939. There weren't any drastic engine or driveline changes, but the H was more comfortable and easier to operate than the original, and more economical to run as well. A five-speed transmission allowed up to nearly 16mph, as long as you also specified pneumatic tires, and there was a choice of a gasoline or kerosene-fueled engine. According to the Nebraska tests, carried out in October 1939, the gasoline version delivered a maximum 24.3hp at the belt and 19.8hp at the drawbar, which made it only slightly more powerful than its predecessor on belt work, but with a 25% boost at the drawbar.

The H was the mid-range Farmall, sandwiched between the smaller A and larger M. It was also a better seller than either of them, with over 300,000 sold between 1939 and 1953. The H was actually International's best selling tractor until the late 1940s, when the M overtook it. Badge-wise, the McCormick name was downgraded to small print and Deering had disappeared altogether—only one name mattered, and that was Farmall.

Left: Better comfort and ergonomics were advantages of the H over its predecessor—not the big coil spring supporting the seat (more effective than a bar of metal), the car-style gear change knob, and hand throttle control.

Above: The Farmall H was introduced just as many original Farmalls were nearing the end of their lives—good timing. It was IH's best seller for several years.

Left: Breakthrough. The Farmall Super MTA of 1954 provided another milestone in tractor design. Its "Torque Amplifier" was a 2-speed planetary gearbox, which doubled the ratios of the standard transmission to 10-speeds. Better still, you could shift between the two ranges without stopping. Within a few years, every major tractor manufacturer had followed IH's lead.

Right: New for 1955 was the numbered series, 100 to 600. This 300 and the 400 had streamlined hoods and "Touch Control" hydraulics. A "Fast Hitch" hydraulic hitch was optional.

Left: The Farmall B was simply a two-row version of the little A, with twin-wheel tricycle front end and an extended left rear axle, though this British-built BMD has a different arrangement. The standard B allowed for a rear axle width of 64 to 92in, though even this wasn't enough for some farmers—the C which replaced it in 1948 gave more rear tread adjustment, on sliding axles.

Opposite page: Farmall 560, a cautionary tale. In the 1950s, while the tractor power race was in full swing, IH concentrated on its smaller tractors. It woke up to this mistake late in the decade, and pushed the 560 into production in 1958. The 60hp six-cylinder engine was fine, but it was mated to the old Farmall transmission, originally designed for the 33hp M 20 years before. It was a disaster, so unreliable that many 560s had to be recalled at great expense. Can it be any coincidence that 1958 was also the year that John Deere overtook IH as the dominant American tractor manufacturer? The 560, according to author P. W. Ertel, was, "easily the biggest blunder in company history."

Left: IH's two-point "Fast Hitch," but still no three-point hitch for the 560.

Left: By 1976, when this 1486 left the factory, all IHs were badged Internationals, and the Farmall name was finally dropped. The pure row-crop tractor had been squeezed out of existence, and the Farmall name went with it. This 1486 was the second most powerful of that new '76 range, with a 145hp turbo diesel. Unusually, this example doesn't have twin rear wheels or front-wheel-assist, normally essential to make use of that much power.

Right: Reflecting modern tractor developments, the 86-series offered hydrostatic drive as well as a conventional transmission. More significant still was the optional cab, with air conditioning, radio, tape player, and hydraulically-mounted seat all available.

Left: For the '70s, International unveiled the 66 series, which featured an all-new cab and ranged from the 80hp 766 to the 133hp 1466. There was a 145hp V8 1468, as well.

Above: Late '60s Farmall 1256, one of the last to carry the Farmall badge. Powered by a 407ci turbocharged six-cylinder diesel, with twin wheels, and optional cab on this one.

John Deere

Left: John Deere stuck with a twin-cylinder format for nearly half a century, and thereby made its fortune.

Classic Profile

It may not look like a landmark machine, but it is. The Waterloo Boy not only helped launched John Deere into the tractor business, but laid down its classic twin-cylinder layout into the bargain. Deere & Co, founded in 1837 by blacksmith John Deere to make plough shares, had already made one attempt to build its first tractor. But the four-cylinder Dain-Deere, designed by Board member John Dain, was complex and untried—it also cost $1,700, when a Fordson was $1,000 less. It was not pursued.

A solution arrived in the form of the Waterloo Gasoline Engine Company, which was already producing a practical and proven machine, the Waterloo Boy. John Deere bought up the whole caboodle, and got itself a ready-made entrée to the business. Of course, even if it failed, the Waterloo Boy's proven twin-cylinder engine gave John Deere a potential way into the stationary engine market.

The Waterloo Boy was quite unique, with its twin-cylinder engine mounted horizontally in the three-

Right: The Waterloo Boy's twin-cylinder format wasn't particularly advanced, but laid the basis for several generations of successful John Deeres. The 1914 Model R produced 25 belt hp from its 333ci power unit, though more power was to come.

SPECIFICATIONS (1920)	
Engine	464ci (7.3 liter)
Bore x stroke	6.5 x 7.0in
Transmission	2-speed
Horsepower	16/25hp
Fuel	3.80 galls/hr
Weight	6,135lb (2,789kg)

wheeler chassis—other tractors had their twin or four-cylinder engines mounted vertically.

The Model R Waterloo Boy produced when John Deere took over was powered by a 333ci motor rated at 750rpm, and producing 12 drawbar hp, 25hp at the belt. There was nothing revolutionary about it, but this simple twin-cylinder tractor was practical, and it worked. It didn't even have a gearbox at first—the Model R had a single forward ratio, though the Model N of 1917 added a two-speed transmission, plus more power from a bigger engine. In fact, that leads to an easy point of recognition between the two—the N has a huge drive gear for each rear wheel, nearly as large as the wheel itself.

By 1923, the Waterloo Boy was looking big and clumsy, but with hindsight its achievement was clear—it laid down the twin-cylinder format that every John Deere was to follow for the next four decades. Quite a legacy in the history of tractor manufacturing.

Right: John Deere had trouble developing its own tractor, so bought the Waterloo Boy concern, lock, stock and barrel. It was the best move they ever made.

Below: A big sturdy crankcase housed the crankshaft for the Waterloo Boy's rearward facing cylinders—on all subsequent John Deeres, the cylinders faced forward. The belt drive in the right of this picture is the cooling fan drive.

Above: Like most early tractors, the Waterloo Boy ran on kerosene, though the transverse-mounted radiator, remote from the engine, was less usual.

Left: John Deere's first GP, originally named the C, powered by a 311ci version of the twin-cylinder side-valve engine. A mechanical implement lift used engine power.

Above: GP WT—WT stood for "Wide Tread," which is self-explanatory, plus a narrower hood and radiator to improve visibility. The 76in rear tread width allowed it to straddle two rows.

Classic Profile

The Waterloo Boy had given John Deere a toe-hold in the tractor market, but it was sadly outdated in many ways. That could have been the end of John Deere as a maker of farm tractors, except that Leon Clausen (later to head up JI Case), insisted that the company build something a little more contemporary.

Not too contemporary, as the Model D launched in 1923 retained the Waterloo Boy's engine format, with twin horizontal cylinders. But there the resemblance ended. Any doubts the company can have had must have soon disappeared. One thousand Ds were sold in its first year, and ten times that number in 1928. It remained in production until 1953, by which time over 160,000 had been built. On the face of it, this success was hard to understand—compared to four-cylinder rivals with three- or four-speed transmissions, the two-speed John Deere looked outdated. Its twin-cylinder motor didn't have the power and smoothness of a four, it was lower revving, and less

Right: Apart from its abortive first attempt (the four-cylinder Dain-Deere), this was the first tractor to bear the John Deere name. Could they have guessed it would still be a major force 80 years later?

SPECIFICATIONS

Engine	465ci (7.3 liter)
Transmission	2-speed
Horsepower	15/27hp
Weight	4,403lb (2,001kg)
Numbers built	161,270
Wheels & tires	28x5/46x12
Price new	$1,125

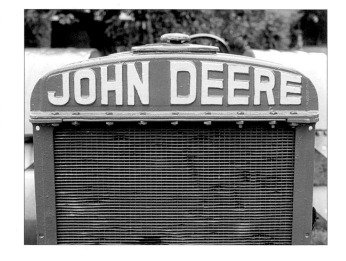

Above: The D was John Deere's most significant tractor. It established them as a major manufacturer, and built up a following for the twin-cylinder format.

sophisticated. On the other hand, it was simpler (a big plus for farmers buying their first tractor), and extremely reliable. It was easier to service and repair than more complex rivals. And its low-revving torquey nature enabled it to slog through hard work. This was the "Johnny Popper" that two generations of American farmers loved.

In any case, if you looked beyond the basic engine format, the John Deere D wasn't at all outdated. Its chassis-less construction meant it was lighter and more compact than the Waterloo Boy, and well up to modern standards. The driveline was fully enclosed, with an oil-bath chain. It was more maneuverable as well, and this all added up to a tractor that was in tune with what farmers needed, and felt comfortable with. They didn't mind a little vibration, but they did appreciate simplicity and reliability. There were updates, too. Power gradually rose from 27hp to 42hp by the final year. A three-speed transmission arrived in 1935, and styled bodywork the following year. The D established John Deere as a major tractor manufacturer.

Above: Rubber pads on the wheels of this D, to improve comfort for the driver on tarmac roads.

Far left: Everything on the D was simple, straightforward and easy to replace.

Right: More comfortable than it looked, with space to stand or sit on the D when you tired of either.

Left: The GP was John Deere's row-crop version of the D, with a smaller 311ci engine. It was still a twin, of course, but it faced a tough time against rivals like the Farmall.

Above: John Deere A, launched in 1934 with overhead valve engine, hydraulic power lift, and power take-off. This is the AOS orchard version.

Above: Neat little John Deere B, "two-thirds the size in power and weight," of the 24hp A, according to company publicity. It was a genuine miniature of the bigger John Deeres.

Right: This is a B too, albeit in pre-Henry Dreyfuss styling, here in BR standard front form. There were many variations on the theme, all with a 149ci twin-cylinder engine.

Left: John Deere AO, designed for orchard work with rounded bodywork and low profile, to avoid snagging branches. Like all As, it offered an hydraulic power lift.

Above: Model A with adjustable wide front axle. All John Deere As used a 206ci version of the twin-cylinder engine, giving 24 belt hp and 16hp at the drawbar.

Below: Freshly serviced and ready to go. A major reason why John Deeres were so popular with farmers was their very simplicity and ease of maintenance. This counted for much at a time when most farmers were still on their first tractor—now, three generations have been weaned on them.

Above: The GP was John Deere's response to the phenomenal success of the Farmall, still based on the twin-cylinder layout. After some years of development, it turned into a useful machine, with high crop clearance, adjustable rear tread on splined axles, and very good visibility. it was powerful enough to haul a four-row cultivator.

Right: John Deere hired industrial stylist Henry Dreyfuss in the mid-1930s, just as International had gone to Raymond Lowey. Dreyfuss' boxy lines suited the workaday John Deere.

70

Power Steering

John Deere might have been one of the first with a diesel tractor in 1949, but the big five-plough R and the 80 that replaced it weren't suitable for row-crop work. Meanwhile, rivals like the Farmall, Massey-Harris, Oliver, and Minneapolis Moline all brought out their own row-crop diesels—John Deere needed to catch up, and fast. It did.

Using experience of the big R, it produced a 380ci diesel version of the classic twin-cylinder motor, not much smaller than that of the 416ci R, but the following year that was boosted to 475ci. Such big cylinders on a high compression diesel were evidently thought too much for an electric start system (though the R had used one). Instead, both 70 and 80 were equipped with a four-cylinder gasoline donkey engine, whose job it was to warm up the big diesel (using exhaust gas heat), before attempting to start it. Interestingly, this high-revving little motor was John Deere's first shot at building a tractor engine with more than two cylinders (a four-cylinder hay baler had been built the year before).

Below: John Deere's 70 Diesel set new standards for fuel economy. Thirty years later, only four other tractors tested by Nebraska had surpassed it, and two of those were John Deeres.

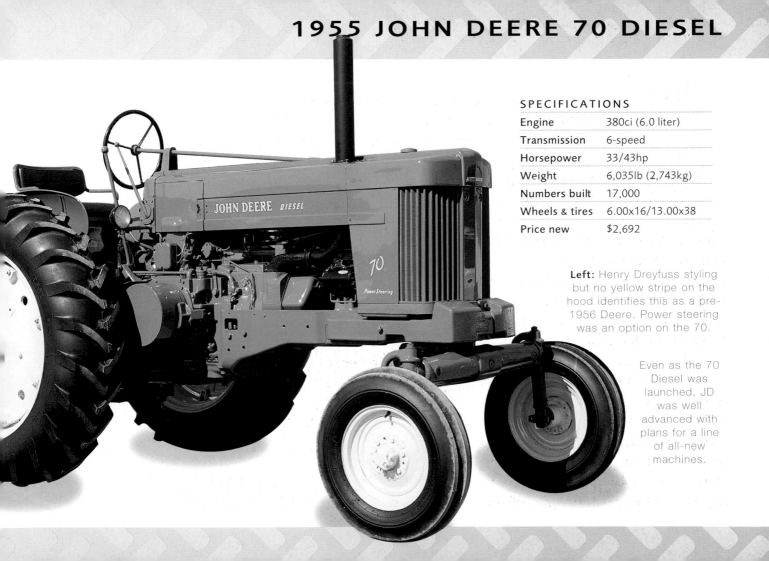

1955 JOHN DEERE 70 DIESEL

SPECIFICATIONS

Engine	380ci (6.0 liter)
Transmission	6-speed
Horsepower	33/43hp
Weight	6,035lb (2,743kg)
Numbers built	17,000
Wheels & tires	6.00x16/13.00x38
Price new	$2,692

Left: Henry Dreyfuss styling but no yellow stripe on the hood identifies this as a pre-1956 Deere. Power steering was an option on the 70.

Even as the 70 Diesel was launched, JD was well advanced with plans for a line of all-new machines.

It worked, which was probably just as well, because the all-new range of four- and six-cylinder Deeres was already on the drawing board.

But for the time being, the 70 Diesel would do well enough. Not only was it powerful enough to be competitive against top-class rivals, with over 42hp at the belt, but it turned out to be the most fuel-efficient tractor of its time. Fuel economy was 17.74 horsepower-hours per gallon, a record figure that stood unbeaten for ten years. It was this outstanding performance of the 70 that brought diesel power into the mainstream of row-crop tractors—the economics just could not be argued with.

Right: Still no three-point hitch for the 70 Diesel when it was launched in 1954. That came with the 720 that replaced it two years later. The 20-series were the first John Deeres with that yellow stripe.

Below: With more sophistication came more information for the driver, dominated by the engine rpm/total hours counter in the center.

Above: Diesel fuel had to be carefully filtered before delivery, but in almost every other way, it was a more reliable system then gasoline, needing no ignition system.

Left and above: Despite its commitment to the twin-cylinder mantra, John Deere was far from being left behind. The R was one of the first mainstream diesel tractors on the market, in 1949, and was also JD's biggest model yet—a five-plough machine producing 51hp.

Below: Power steering and electric start on the John Deere 60, reflecting a more demanding post-war world. It was launched in 1952 as a direct replacement for the Model A, with the addition of a live power take-off and some minor engine changes. The A had already been fully updated five years before, with standard electric start and stamped steel chassis.

Right: In 1955, JD replaced the R diesel with the 80—a bigger 475ci version of the twin-cylinder diesel, and a six-speed transmission.

Left: This 420 appeared in 1956, the first of the company's 20-series. It was basically a more powerful version of the 40, which itself had replaced the 101ci M three years before.

Above: A new idea for 1959, the 8010 was one of the first four-wheel-drive articulated tractors. It was powered by a six-cylinder GM two-stroke diesel of 215hp.

4010 DIESEL

You had to hand it to John Deere. They had persisted with two cylinders where every rival had gone to four, and made a success of it—even in the late 1950s, the "Johnny Popper" twin retained a huge following. But by then, it was time for a change: the twins were getting left behind in the horsepower race, they needed heavy duty transmissions, and in any case, the tooling was simply wearing out. Some manufacturers might have rushed out a four-cylinder motor, or bought someone else's. Not John Deere. Instead, they took their time—a four-cylinder mockup was built as early as

SPECIFICATIONS

Engine	380ci (6.0 liter)
Transmission	16-speed
Horsepower	74/84hp
Weight	6,525lb (2,966kg)
Numbers built	36,500
Wheels & tires	6.00x16/15.50x38
Price new	$4,294.50

1950, and once the decision was taken to go ahead with an all-new generation of tractors, it was seven years before they were unveiled. In great secrecy, a small team of engineers was assembled in a disused grocery store in Waterloo, not far from the John Deere factory. As the project expanded, different parts were farmed out to different workshops, so that no one could put two and two together. When they were ready, the prototypes were disguised

Left: Fingertip control, thanks to modern ergonomic design: engine, transmission, hydraulic, and PTO controls were all accessible without bending or stretching.

Right: Range-topping 4010 Diesel was lighter than the 730 it replaced, but had 50% more power and put only 18% on the price.

1963 JOHN DEERE 4010

Only one thing was carried over from the old
John Deeres—the color scheme.

with false body panels, and the company bought a 684-acre farm a few miles south west of Waterloo, where they could be tested away from prying eyes.

When the result was finally launched on August 30th, 1960, JD revealed a complete range of four- and six-cylinder machines, the 1010, 2010, 3010, and 4010. Totally new from stem to stern, they had cost $50 million to develop—the company's future rested squarely on their success or failure.

John Deere's engineers had experimented with V6 and V8 engines, but these restricted visibility and were expensive to build. Instead, there were relatively conventional gasoline and diesel in-line units, ranging from 36hp to 84hp. All the New Generation tractors placed a new emphasis on driver comfort—so while Henry Dreyfuss had again taken care of the styling and driver station, an orthopedic doctor was consulted to ge t the ergonomics right. Adjustable suspension for the seat, and a roomy platform with easy access helped make this the most comfortable John Deere ever. The company had already pioneered the use of hydraulic implement lifts, but on the 10-series tractors one integral system provided hydraulics for the power steering, three-point hitch, power brakes, and differential lock. John Deere had transformed itself, and how.

Above: Serial number for 1963 4010, but even then, the replacement 20-series was imminent. John Deere was determined not to get left behind again.

Left: The 4010 shared many features with the 3010, notably the new hydraulic system, but added an 84hp six-cylinder engine.

Right: New three-point hitch used the lower links to sense draft. Using the top link (the norm on three-point hitches), had its limits on heavier implements.

Left: This was the final incarnation of the twin-cylinder John Deeres. The 435 was introduced in reaction to Massey-Ferguson's small diesel utility tractor. It was replaced by the 1010 New Generation.

Below: 3010 was similar to the 4010, but with a 254ci four-cylinder engine in place of the 380ci six.

Left: 4320 Diesel, one of the last New Generation. Apart from some reliability problems with the small 1010 and 2010, the New Generation was a complete success, but many updates appeared over the next decade. A 121hp 5010 joined the range in 1963, and the turbocharged 4520 five years later, with hydrostatic front wheel assist. The 4320 shown here was a powered-up 115hp version of the basic 4020. Meanwhile, both 4520 and 5020 acquired intercoolers, boosting power to 135hp and 175hp respectively.

Right: A Sound-Gard cab was the big new feature on the Generation II John Deeres, which were unveiled in August 1972. It set new standards for quietness, being attached to the tractor via four rubber mounts. Seat, floor and controls were all integral to the cab, and the big glass area pointed to the future of cab design.

By 1972, the New Generation was new no longer. Despite the updates, it was now twelve years since John Deere's radical new range was unveiled at the Dallas launch. A replacement appeared in August 1972, the "Generation II," instantly recognizable by its new styling, and the 30-series suffix.

The most obvious change was the Sound-Gard cab as detailed on the previous page. This also had standard four-post ROPS (roll-over protection), and although the cab cost extra, three out of four 30-series customers opted for it.

Above: The 126hp 4430 was really reaching the limits of pure two-wheel-drive.

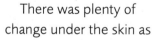

There was plenty of change under the skin as well, with a simplified, rejigged collection of mid-range tractors. The 4430 on these pages was the second largest of four, a replacement for the 4320. But not all the 30-series were built in America. John Deere now had plants all over the world, and from the mid-'60s began sourcing all its smaller tractors from the Mannheim factory in Germany. The US-made range

Far left: Not a speed range change, but the speed with which the hydraulics operated.

1973 JOHN DEERE 4430

SPECIFICATIONS

Engine	404ci (6.4 liter)
Transmission	16-speed
Horsepower	105/126hp
Weight	11,350lb (5,159kg)
Numbers built	61,960
Wheels & tires	10.00x16/18.4x38
Price new	$22,575

Left: Twin wheels on this 4430—most buyers preferred this system to the more expensive front-wheel-assist or four-wheel-drive.

New boxy styling for the 30-series, which raised a few eyebrows at first.

kicked off with the 80hp 4030, alongside the 100hp 4230, 125hp 4430, and 150hp 4630. The three larger ones all used the same 404ci six-cylinder diesel, in naturally aspirated, turbo, and turbo-intercooled form respectively. The 4430 was available as a row-crop, standard, or hi-crop—their combined sales made it the most popular 30 series.

All of these tractors had the 8-speed Syncro-Range transmission as standard, with Perma-Clutch, a new wet hydraulically-controlled clutch. A 16-speed Quad-Range transmission was optional, with a built-in Hi-Lo no-clutch change. Power Shift was on the options list as well, not to mention front-wheel-assist. The latter, especially on the 4430, might have been thought highly desirable on a 120hp+ tractor, but few buyers opted to pay for it—twin rear wheels was a simpler and more common alternative.

If none of these tractors were big and powerful enough, John Deere also updated its four-wheel-drive articulated range. These had been part of the line-up since the original 8010 of 1960, and now comprised the 215hp 8430 and 275hp 8630. Now covering everything from mini lawn tractors to these prairie monsters, the John Deere line up looked more secure than ever before.

Above: Sound Gard cab set new standards for quiet and cleanliness—it was rubber mounted, pressurized to keep dust out, and air conditioning was among the options. A huge glass area ensured that the roll-over protection didn't interfere with all-round visibility.

1976 JOHN DEERE 4430

Left: The 4430 was the best selling of the US-built 30-series tractors, a natural successor to the 4320, 70 Diesel, and, for that matter, the Model D.

Above: A six-cylinder diesel had become the standard power unit for mid-size tractors, in this case turbocharged. Lubrication and cooling systems were boosted to match.

Left: Through the difficult 1980s and '90s, John Deere kept up to date with a stream of new models. This is the 100hp, Mannheim-built 6400.

Right: By 1982, six turbocharged cylinders weren't always enough. John Deere's response was the V8 8850, turbocharged and intercooled to deliver 300hp at the PTO.

Above: 7810 MFWD John Deere pulling a HX14 Cutter in wheat. This is one of the John Deere company's own publicity shots for this model.

Right: Waterloo-built 7800—the smaller 6000 series was made in Germany. This one topped the range, with a 140hp 487ci power unit, 19-speed powershift, and front-wheel-assist.

Left: 8100, one of a new range of large two-crop tractors launched in 1995. A Caterpillar-style rubber track was optional from '97.

Right: John Deere survived the difficult 1980s and '90s as a worldwide company—this 6400 of the mid-'90s was built in Mannheim, Germany.

Below: PowerQuad transmission on this 1996 6900—four ranges (power shiftable) with 20 forward speeds.

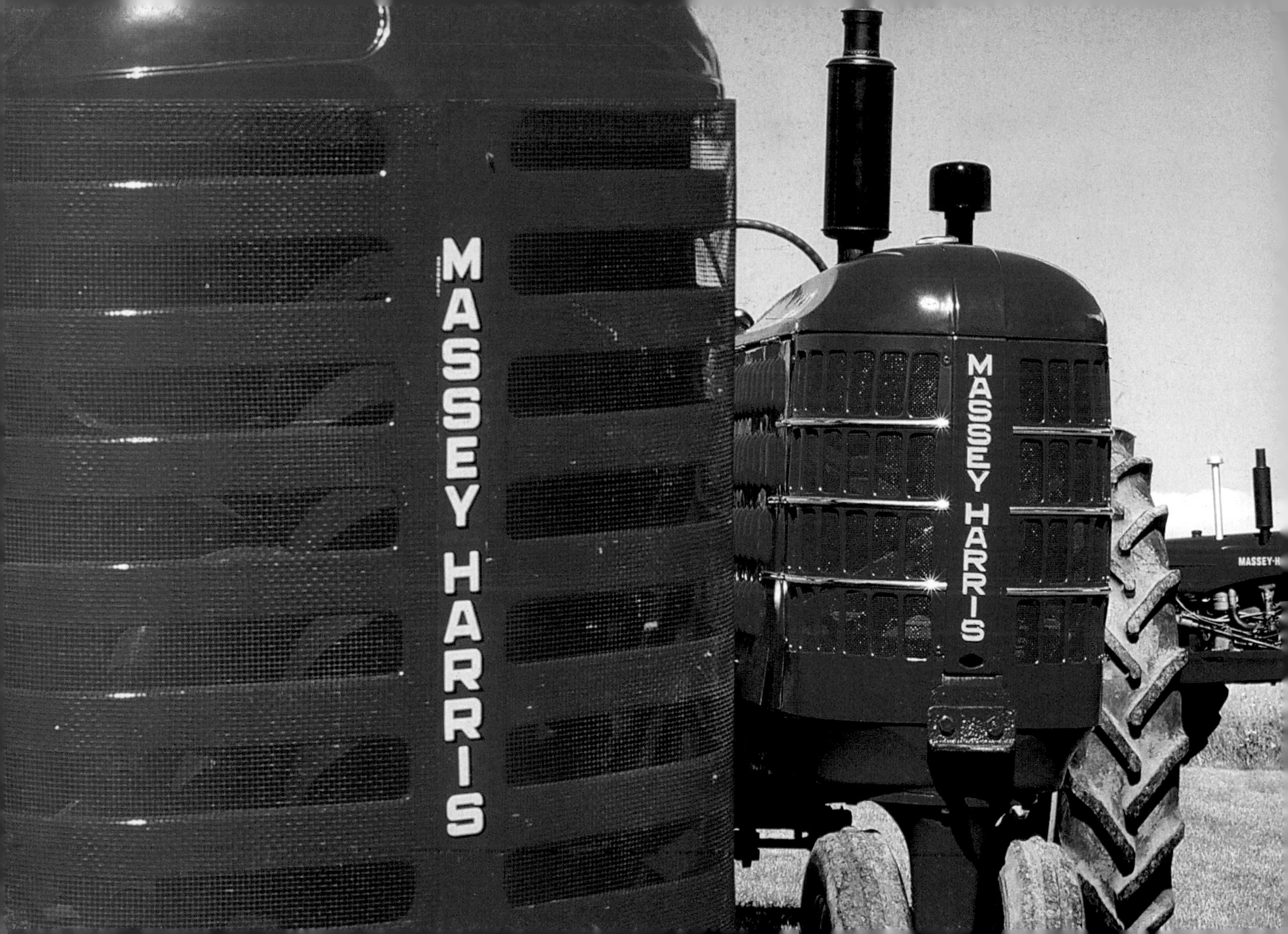
MASSEY HARRIS
MASSEY HARRIS

Massey-Harris

Left: Canada's most famous tractor manufacturer, Massey-Harris, later merged with Ferguson.

Below: The first tractors to bear the Massey-Harris name weren't actually made in Canada, apart from a few. M-H commissioned Parrett Tractors of Chicago to design a range for them—this is the 12/22hp No2 of 1921.

Right: Along with its 12/25 and 15/28 siblings, the No2 Massey-Harris looked outdated and expensive next to a Fordson. M-H had to look elsewhere for its tractors.

Classic Profile

MASSEY-HARRIS

Today, four-wheel-drive and chassis articulation are common features of tractor design—indeed, the big super-tractors never use anything else. But in 1930 they were unheard of; Massey-Harris' General Purpose machine of that year pioneered both. It was a clever piece of design, carefully thought-out, though ultimately a failure. Traction was all on the GP, and Massey-Harris sought to maximize the benefits of four-wheel-drive through what it called "balanced traction." Firstly, the articulated chassis allowed front and rear axles to follow their own path over obstacles, helping to keep all four wheels on the ground. Second, with a forward-mounted engine, the GP's weight was biased towards the front, so the weight of a draft implement tended to equalize the ground pressure on all four wheels. The idea was that such an adaptable tractor could pull drawbar loads over any terrain, cope with row crops in heavy soil and do belt work—hence the name, General Purpose. Rivals like the International

SPECIFICATIONS	
Engine	226ci. (3.6 liter)
Transmission	3-speed
Horsepower	15/22hp
Weight	3,861lb (1,755kg)
Numbers built	c3,500
Wheels & tires	38x8/38x8
Price new	$1,245

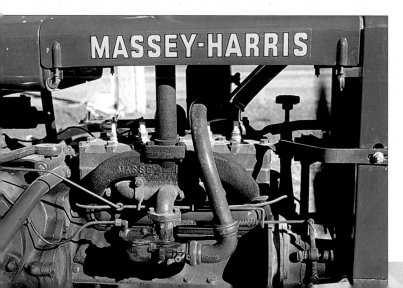

Left: Early GPs used a bought-in Hercules engine, but from 1936 Massey-Harris fitted its own power unit.

1930 MASSEY-HARRIS GP

Right: A modern concept, but the Massey-Harris still used a simple side-valve four-cylinder engine and basic three-speed transmission

Above: According to the serial number, this GP was built during the first year of production.

Four equal-size driven wheels plus articulation made the GP a good wheeled alternative to crawlers.

Farmall had already shown there was a huge market for a machine that could do all of this.

Unfortunately, clever though it was, the Massey-Harris GP didn't fulfill the hopes of its designers. All this technology didn't come cheap—the GP cost over $1,200, substantially more than a Farmall. Its axles couldn't be adjusted for different row crop widths (though they were supplied in six alternative fixed widths by the factory), and the ten foot turning circle also detracted from the easy-to-use nature required of a do-it-all tractor.

So, not a successful general purpose machine, but the GP did find its niche in forestry work, where its excellent traction over difficult terrain was a big advantage—it was the only wheeled tractor of its time which could do the work of a crawler, and was favored by some logging companies.

Left: The early GP used a Hercules engine, but Massey-Harris designed its own oil-bath air cleaner to go with it.

Left: In some ways, the Massey-Harris GP was a concept ahead of its time. Four-wheel-drive and articulation took 30 years to reach the mainstream.

Above: The GP's forward weight bias is obvious here. Its engine was mounted ahead of the front axle, to counteract the weight of a towed implement and thus give even weight distribution over the four driven wheels.

Left: GP with rubber-covered steel wheels. There was also an industrial version with pneumatic or solid rubber tires, plus an airport model with smooth steel wheels.

Above: From the mid-1920s, Massey-Harris distributed Wallis tractors, and in 1928 took over the company altogether. A range was built, all with the distinctive U-shaped chassis.

Below: Early GP. An electric start was fitted later, along with Massey-Harris' own four-cylinder engine, and experiments were made with six-cylinder units. Post-1936 GPs can be identified from their forward sloping hoods.

Right: Alongside the avant garde GP, Massey-Harris sold the more conventional Challenger from 1936. It was an update on the old Wallis 12/20, with overhead valve 248ci engine and adjustable rear track. The Pacemaker was a standard tread version of the same tractor.

Left: The 101 was a striking looking tractor, unlike any previous Massey-Harris. In place of the Wallis U-shape unit chassis, the 101's gearbox and rear axle formed the rear part of the frame, while the engine was supported by a cast-iron subframe.

Left: Nineteen thirty-eight, and the 101 is unveiled, bringing Massey-Harris right up to date. Six-cylinder Chrysler engine, fully enclosed in streamlined styling, with a four-speed transmission.

Right: Under that sleek hood, the 201ci Chrysler six produced 36hp at the belt and 24 at the drawbar. In the Super 101 it was boosted to 218ci.

Left: In 1941, the 81 was Massey-Harris' smallest tractor, a cheaper alternative to the 101 Junior. A 2,700lb lightweight, with a little 124ci Continental providing the power.

Above: The 81 had a two-speed governor (Twin Power) which M-H had pioneered with the Challenger. After the war it was updated as the Model 20, still two-plough.

Left: 102 Junior, which with its162ci Continental engine slotted between the 124ci 101 Junior and the big six-cylinder 101. A 2-3 plough tractor, in row-crop or standard tread.

Below: Post-war, the 33 was a three-plough machine with both diesel and LPG options. The standard 201ci four-cylinder engine produced 28/30hp.

Below: In 1952, this was Massey's latest small tractor. The Mustang replaced the little 22, which could trace its roots back to the pre-war 81. Once again, Continental provided the power (a 140ci four)—the Colt was a 124ci version.

Right: One of the few Massey-Harris tractors to be actually built by M-H in Canada—the Pony. This was a little single-plough machine, powered by a 62ci Continental engine of 8/10hp. It was also assembled in France with Simca or Hanomag diesel power units.

Left: One of the last products of the independent Massey-Harris, and also one of its most successful. The 44 came with four or six-cylinder engines of about the same power.

Above: Successful though it was, the 44 lacked a three-point hitch with draft control, just like every other Massey-Harris. The 1953 merger with Ferguson provided a solution.

9

Minneapolis-Moline

Left: Minneapolis-Moline mixed many conventional tractors with some astonishing innovations.

Above: Before the 1929 merger that produced Minneapolis-Moline, Minneapolis Steel & Machinery built tractors under the Twin City brand, such as this giant 60-90 of 1913.

Right: Minneapolis' early big tractors sold in small numbers, but it also built machines for Bull and Jl Case. Not until 1919 did it offer a truly modern tractor of its own.

Left: Another partner in the 1929 merger was the Minneapolis Threshing Company, which produced this four-cylinder cross-engined tractor in 1920.

Above: The car-like, fully-enclosed 16/30 was a departure for Twin City in 1917. Electric lights and starter were optional, though the 16/30 was not a great success.

Below: There's nothing new under the sun, as this Twin City 12/20 demonstrates. Launched in 1919, its four-cylinder engine boasted four valves per cylinder, decades before this became a feature of high performance cars. It was also a turning point for Minneapolis, from old-school heavyweight tractors to lighter more advanced machines. The 12/20 cost twice as much as a Fordson, but still sold very well.

Above: So successful was the 12/20 concept of unit construction and 16-valve engine, that it spawned the bigger 17/28, then this, the 21/32, in 1929.

Left: The 21/32 still only had a two-speed transmission (a three-speed came later), but did have a gear-driven oil pump and enclosed lubrication for both gearbox and steering gear. Renamed the FT in 1932, it carried on, 13 years after the original concept was introduced.

Left: Minneapolis-Moline MT, a row-crop version of the KT. Both used a 284ci four-cylinder gasoline engine producing 18hp at the drawbar, making the MT a 3/4-plough tractor.

Above: Great leap forward. Minneapolis-Moline caught the opposition with this 335 and the 445 in 1955: ten-speed transmission, three-point hitch, and live PTO.

Above: All-new for 1936, the Z-series included a number of innovations, such as a five-speed gearbox, stylish sheet metal and a new 185ci engine designed for easy maintenance.

Right: The Z came in a number of forms, including ZTU (above) and ZTS (right); "Z" was the series, "T" stood for tractor, and "U" or "S" for row-crop or standard tread.

Above: UTS—"U" for M-M's big four-plough tractor, and "S" for the standard tread front end. UTU was the row-crop version, both with the same 283ci four-cylinder engine.

Right: U-series, here in UTC high-clearance sugar cane form. The U was a much bigger four-plough version of the little Z, with a 283ci engine giving 65% more power.

Classic Profile

Did someone at Minneapolis-Moline have a crystal ball? If they had, maybe they wouldn't have produced the UDLX Comfortractor. It was wildly ahead of its time but too expensive to sell in more than small numbers. With its fully enclosed cab, heater, and radio, the Comfortractor was better appointed than some cars of the time. It even had a road speed of 40mph, thanks to Minneapolis-Moline's now trademark five-speed transmission. In this tractor, so the theory went, you could plough all

SPECIFICATIONS

Engine	283ci (4.5 liter)
Transmission	5-speed
Horsepower	31/38hp
Weight	6,000lb (2,727kg)
Numbers built	c150
Wheels & tires	7.50x16/12.75x32
Price new	$2,155

Left: You didn't clamber aboard the Comfortractor, as you did its rivals, but stepped in decorously, via the large rear-mounted door.

Right: Under its shapely skin, the UDLX made use of the same 283ci four-cylinder engine as other U-series Minneapolis-Molines.

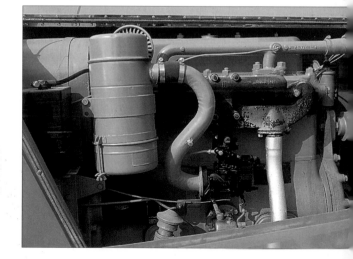

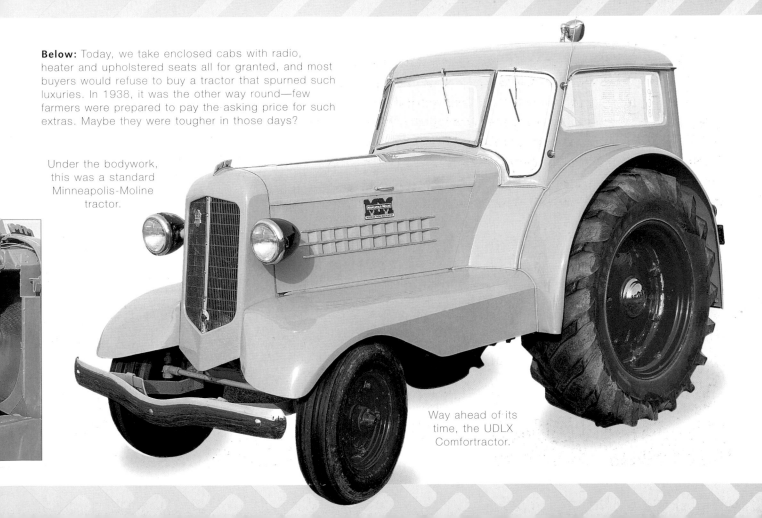

Below: Today, we take enclosed cabs with radio, heater and upholstered seats all for granted, and most buyers would refuse to buy a tractor that spurned such luxuries. In 1938, it was the other way round—few farmers were prepared to pay the asking price for such extras. Maybe they were tougher in those days?

Under the bodywork, this was a standard Minneapolis-Moline tractor.

Way ahead of its time, the UDLX Comfortractor.

Classic Profile

Above: If the Comfortractor was rare, this open top version was rarer still. It's hard to see why anyone would pay the extra and not have the benefit of that cab. Still, you could hardly choose a more elegant means of farm work!

Left: Detailing underlined that this tractor was designed to be used on the road—automobile components were used all over it.

Right: Tractor or motor car? From some angles it was difficult to tell the difference!

day, in great comfort, then drive straight into town—the farmer's wife could come too, on the upholstered passenger seat. It looked good as well, with a chrome front bumper and a healthy slice of 1930s style. Nor was it just a cosmetic exercise—details like safety glass, a defroster, and foot accelerator made this a practical road vehicle, though that cocoon-like cab did restrict vision a little in the field.

At $2,155, the Comfortractor was simply too expensive to attract farmers still recovering from a long depression, and only 150 or so were built.

Left: Despite the poor sales of the Comfortractor, Minneapolis-Moline hadn't given up on the factory cab, and offered it as an option on certain models—even this little two-plough RTU.

Above: Unveiled in 1939, the R was a smaller version of the Z, with a 165ci employing the same easy-maintenance features. The crankshaft could be checked without even draining the sump.

Right: Full fenders for this RTS standard tread version. RTU was the row-crop, RTE had a wide-front adjustable axle, and RTI was the industrial version.

Below: By the late 1930s, farmers were once again demanding more power. Minneapolis-Moline obliged with the GT in 1938. The company's biggest tractor since the old Twin City days, it used a 403ci engine based on that of the old 21/32, but with a lot more power—36hp at the drawbar was enough to make this a five-plough machine, with 49hp at the belt.

Above: In 1947, GTA became GTB, and power was boosted to 39 drawbar horsepower, 51 at the belt. Also that year came the LPG powered GTC, still based on the same 403ci four-cylinder engine.

Right: Red grille denotes 1942 GT, which later that year became the GTA and went back to a yellow grille. There were apparently no other changes.

1943 ZTX • • MIL 224

Left: LPG—Liquid Petroleum Gas—was an increasingly common option for tractor buyers in the 1950s, an era when different fuels were fighting for supremacy: kerosene was outmoded, gasoline more powerful but more thirsty, and diesel cheaper to run. LPG was a cheaper fuel, but not as efficient as diesel. This Minneapolis-Moline GTC was a factory conversion of the GTB, available from 1951 to '53.

Right: Ultra-high ground clearance was the name of the game for sugar cane work, which became something of a Minneapolis-Moline specialty.

Left: During WWII, Minneapolis-Moline produced many military vehicles, including a prototype tractor-based jeep for the US Navy. This ZTX was designed for short haulage of heavy loads, using the running gear of the Z series, with five-speed transmission, and steel cab from the R series.

Left: In the late 1940s, Minneapolis-Moline was lacking a small single-plough tractor to compete with the popular Allis-Chalmers G. So it bought the BF Avery Company in 1950, which already built a suitable machine.

Below: Rebadged, and painted in M-M Prarie Gold, the little Avery became the Minneapolis-Moline V, powered by a 65ci Hercules engine. But it was not a success, and only remained in the line-up for a couple of years.

Left: Diesel power came to the big GTB in 1953. This updated GBD of 1955 used a six-cylinder engine of 425ci. It cost $4320 new and weighed in at 7,400lb.

Above: M-5, 1960. While many manufacturers were moving to six cylinders, M-M stuck with four, though the M-5 did have live hydraulics, a two-range transmission, and power steering.

Left: Minneapolis-Moline was a keen proponent of four-wheel-drive, building the industry's first sub-100hp 4x4 machine, the M504, in 1962. So when the big Gvi (which had replaced the GB series) was dropped in 1963, its successor had a four-wheel-drive option from the start. The G706 was powered by a 504ci six-cylinder engine in diesel or LPG form—significantly, there was no gasoline option. G705 was the two-wheel-drive version. These were the last new M-Ms before the takeover by White Motors.

Right: Somewhere within all that machinery is a Minneapolis-Moline GTA. It's a rice field special, which required ultra high ground clearance and very large, low impact tires.

Classic Profile

If the Minneapolis-Moline M670 qualifies as a landmark, it's a sad one, certainly for M-M enthusiasts. It was one of the last genuine M-Ms, before the company's new owner White Motors began to badge-engineer the name out of existence. From the mid-1950s, Minneapolis-Moline had some hard thinking to do. As one of America's smaller manufacturers, it was more vulnerable to the ups and downs of the marketplace than, say, John Deere. Worse still, its tractor line was starting to look outdated, thanks in part to its concentration on the odd-looking Uni-Tractor, and it had no small tractor to offer at all.

All was not lost however, for in 1955 the company unveiled two thoroughly modern new machines, the 30hp 335 and 38hp 445. Both had two-range Ampli-Torc transmissions, with a Ferguson type three-point hitch and live PTO on the options list. That still left the ageing U-series, now not short of 20 years old, and in 1957 M-M replaced it with the 5-Star, though it was still powered by the U's 283ci engine (enlarged to 336ci in diesel form). That same engine, which had been launched in 1938, went on to survive several model

SPECIFICATIONS

Engine	336ci (5.2 liter)
Bore x stroke	4.63 x 5.0in
Transmission	10-speed
Horsepower	62/73hp
Fuel	9.89hp/hr per gallon
Weight	7,395lb (3,361kg)

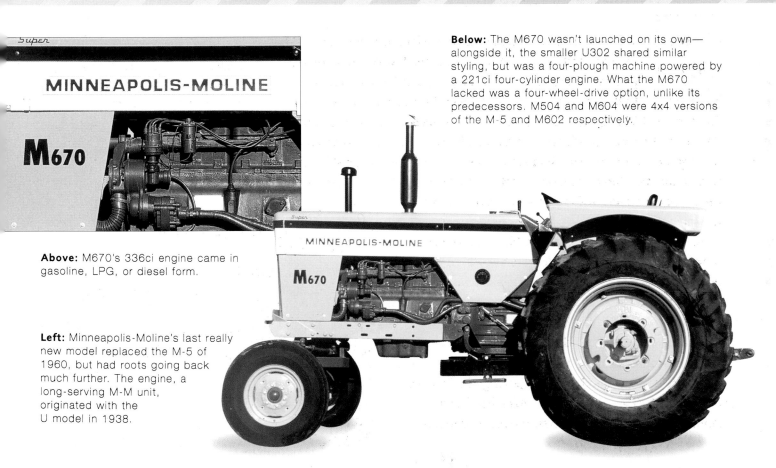

Above: M670's 336ci engine came in gasoline, LPG, or diesel form.

Left: Minneapolis-Moline's last really new model replaced the M-5 of 1960, but had roots going back much further. The engine, a long-serving M-M unit, originated with the U model in 1938.

Below: The M670 wasn't launched on its own—alongside it, the smaller U302 shared similar styling, but was a four-plough machine powered by a 221ci four-cylinder engine. What the M670 lacked was a four-wheel-drive option, unlike its predecessors. M504 and M604 were 4x4 versions of the M-5 and M602 respectively.

changes; the M5, four-wheel-drive M504, M602, and finally this, the M670.

One of M-M's last new tractors, the M670 appeared in 1964 alongside the smaller U302. The venerable four-cylinder engine now came in the 336ci size whether one opted for gasoline, diesel, or LPG power. With squared-off styling to match the big G series, it did look up to date, and the multi-speed transmission and three-point hitch backed that up. But it was something of a swansong for M-M, now being run down as a subsidiary of White Motors. The M670 was dropped in 1970, replaced by a Fiat, which White repainted in M-M colours and badged the G450. But it didn't last long—White dropped the Minneapolis-Moline name altogether in 1974.

Right: Brave attempt. Despite being relatively small, Minneapolis-Moline did a good job of keeping its tractor line up to date between the mid-1950s and mid-'60s. The M670 was a case in point, but it turned out to be the company's last genuinely new machine—after six years, it was replaced by a badge-engineered Fiat.

Below: Squared-off industrial styling was certainly up to date in 1964—it echoed that of the big Allis-Chalmers D21, which had been launched the previous year. Unlike the big Allis, the M670 set no new size or power records, though it was a thoroughly modern and competitive small tractor.

Above: A three-point hitch helped make the M670 competitive, though this was now a minimum requirement. Likewise Ampli-Torc, the two-speed planetary gearbox, that doubled the forward ratios to ten. The M670 was up with the trends, but not ahead of them.

Left: Not really a Minneapolis-Moline at all, but White's 1989 American, painted in M-M colors in an attempt to appeal to folk-memory. It also came in Oliver green and Cockshutt red.

Below: The last big row-crop tractors from M-M were the G1000 series, seen here in LPG-powered G1050 form. The Minneapolis-Moline name was dropped by White in 1974.

10

Other Makes

Classic Profile

Advance Rumely OilPulls were the definitive big, heavy, kerosene-burning tractors of the early days. So well developed were they, that the OilPulls outlived all rivals, still selling long after the new breed of light, nimble gasoline tractors had become dominant—the last one was built as late as 1931.

The secret of the OilPull's success was its huge twin cylinder engine (654ci on this 16-30, which was by no means the largest on offer). The hot-

SPECIFICATIONS

Engine	654ci (10.3 liter)
Transmission	2-speed
Horsepower	16/30hp
Weight	9,506lb (4,321kg)
Numbers built	13,553
Wheels & tires	40x7/56x18
Price new	$2,600

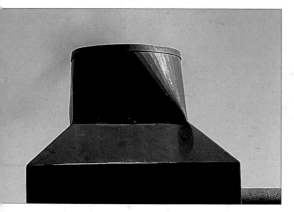

Above: Not steam, but kerosene. In some ways, the OilPull and others like it were stepping stones between steam traction engines and the modern gasoline tractor.

Right: It looks fast, but the OilPull's steam engine-like twin cylinder unit turned over at a stately 530rpm. With its smooth, torquey delivery (just like a steam engine), it was a favorite for belt work.

Steam traction engine, or gasoline tractor? The OilPull was neither.

Above: 16/30 was the mid-range OilPull, and the most popular. Others were the 12/20 and giant 30/60.

running cylinders would run on cheap fuels like kerosene—a water atomizing carburetor prevented them getting too hot, and pre-igniting. The engine wasn't cooled by water, but oil—hence the name.

Instead of a radiator, the OilPulls had a large cooling tower, one of the biggest components on the tractor. With kerosene smoke puffing out of the tower, and the OilPull's traction engine looks, it could often be mistaken for a steam engine. It pulled like one too—Rumely hitched three of its biggest 30/60 OilPulls to a specially constructed 50-bottom plough. They took less than five minutes to work an acre.

Left: Top end of the twin-cylinder engine, showing the valve gear. On the 16/30 Type H shown here, the cylinders were cast as a pair, and the machine used a 180-degree crankshaft for smoother power delivery.

Right: Advance-Rumely of La Porte was later taken over by Allis-Chalmers.

Opposite page: Rumely built OilPulls to the same format for over 20 years. The modern 6C intended to replace them came too late.

SOLD BY
ADVANCE-RUMELY THRESHER CO. Inc.
Incorporated
LA PORTE   INDIANA

MANUFACTURED BY
ADVANCE-RUMELY CO.
Incorporated
LA PORTE   INDIANA
UNITED STATES OF AMERICA

Right: At first, the AGCO Allis tractors were little more than rebadged Deutz-Allis machines. But the US content was gradually increased, though Deutz air-cooled diesel engines were used up to 1996.

Left: Now AGCO Allis is a brand in its own right, covering the mid- to large end of the market. It still shares components with the other AGCO brands, though.

Right: In 2002, just four big companies dominated the tractor industry: John Deere; CNH Global (Case & New Holland); Same Deutz-Fahr of Italy; and AGCO. The AGCO story is remarkable. It was formed in 1990 by a management buy-out of the US arm of Deutz-Allis. Three years later, it bought Massey-Ferguson and White-New Idea; since then, McConnell and Fendt have joined as well. In an industry full of old names, this new one is doing pretty well.

Classic Profile

Two crawlers stand out in Caterpillar history as landmark tractors—the 1986 Challenger and this, the 1921 2 Ton. Why? Because both were aimed, not at Caterpillar's traditional construction market, but at farmers.

The 2 Ton didn't actually start life as a Caterpillar at all, being launched as the Holt T-35. Daniel Best and Benjamin Holt were both pioneer crawler manufacturers, not to mention bitter rivals. So bitter, that at one point they went to court over a patent dispute. But both realized that they couldn't survive the tough 1920s on their own, and merged to form Caterpillar. Together, they built-up the most dominant crawler manufacturer not just in North America, but the whole world.

The pioneering Best and Holts were mostly big, heavy machines, well suited to construction and heavy haulage, but too big for most farmers. Holt's new president Thomas Baxter realized that there was a market in smaller machines, in farming as well as logging and construction. The result was the Holt T-35, soon renamed 2 Ton, to fall in line with the existing Holt 5 Ton and 10 Ton. The smallest in the Holt range, it was powered by an overhead camshaft four-cylinder engine of 251ci. Rated at 1,000rpm, it was originally intended as a 35hp unit, but came out at 15hp drawbar, 25hp at the belt.

Right: The 2 Ton was the first smaller crawler to come out of Caterpillar, though it was really a Holt, designed and put into production before the 1925 merger with Best.

SPECIFICATIONS

Engine	276ci (4.3 liter)
Bore x stroke	4.0 x 5.5in
Transmission	3-speed
Horsepower	15/25hp
Fuel	9.45hp/hr per gallon
Weight	4,040lb (1,818kg)

CATERPILLAR TWO TON

Below: Clean, simple lines, and cheap enough for many farmers to afford—that was what made the 2 Ton a landmark tractor.

Classic Profile

Apart from its unusual overhead cam engine, the little Holt (which became the Caterpillar 2 Ton after the 1925 merger), was unique in locating the transmission behind the rear axle, and in the fact that Holt made every single component itself, apart from the magneto and carburetor.

It was also remarkably cheap—at one point, the 2 Ton was selling for just $375, which made it competitive with wheeled tractors of the same size. Just as the Fordson F introduced a whole generation of farmers to wheeled tractors, the little 2 Ton Caterpillar did the same with crawlers. Little wonder that the names "Caterpillar" and "crawler" are almost interchangeable.

Above: Note tiller steering on the 2 Ton, in place of a steering wheel, and the rubber pads on the tracks, to make road work quieter and more comfortable.

Right: Despite its exotic sounding overhead camshaft (unheard of on a tractor), the 2 Ton failed to make its planned 35hp by a big margin—25hp was all it could muster at Nebraska.

Above: Sound advice. Almost every component of the 2 Ton was made by Holt/Caterpillar itself, apart from the Kingston carburetor and Eisemann magneto.

Below: Surprisingly well padded seat for the 2 Ton driver, who of course would have to get by without the traditional sprung seat of a conventional wheeled tractor.

Below: Caterpillar eventually responded to farmers' demands for a smaller Challenger, more suited to general farm work. The 35 and 45 also had more ground clearance and adjustable tread widths, though they were still big machines, with 210 and 240hp respectively.

Right: Possibly the biggest leap forward ever in crawler technology. Steel-reinforced rubber tracks made crawlers faster, quieter, and much more suitable for field work. Caterpillar's pioneering Challenger was launched in 1986 as a rival to big wheeled tractors.

Crawlers have long played a minor role in agriculture, often favored by growers of specialized crops, or where soil impaction was a real problem. Cletrac—the brand name came from the company name, the Cleveland Tractor Company—built a range of crawlers from 1918, concentrating on the smaller models. This move reflected the ambition of founder Rollin H. White to make tractors suited to agriculture. One of his early innovations, and one which remained unique to Cletrac, was a system which allowed the crawler to be steered with a conventional steering wheel. Normally, crawlers were steered by levers, which declutched the inside track —the problem was that this reduced traction, and in any case not many people were used to steering with levers. Instead, the early Cletracs used a planetary gear set controlled by a brake in each drive cog. This controlled the amount of power going to each track, and allowed smooth steering via a conventional wheel. This

SPECIFICATIONS	
Engine	113ci (1.8 liter)
Transmission	3-speed
Horsepower	11/15hp
Weight	2,950lb (1,341kg)
Numbers built	29,930
Wheels & tires	Steel or rubber tracks
Price new	$2,048

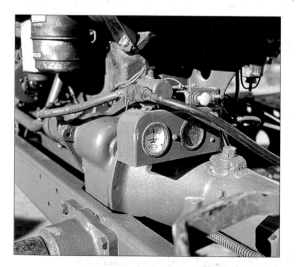

Left: Gauges were tricky to read on the move, even if your top speed was only a slow walking pace.

Right: Small, light and affordable by crawler standards, the Cletrac HG was aimed at small farmers for whom wheeled tractors weren't enough.

Classic Profile

controlled differential later became a common feature in crawler design.
The Cletrac Model F of the early 1920s was specifically aimed at farmers, and was available in high clearance row-crop form. In fact, in 1939 Cletrac introduced a wheeled tractor, the CG. The HG shown here was the crawler version—costing only $2,000 and weighing a ton and half (both modest figures by crawler standards), it brought crawler technology within reach of small farmers.

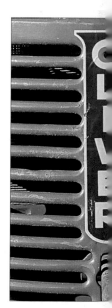

Above left: "68" on the serial number plate denotes tracks 68in apart. A 42in tread was available as well.

Left: A belt pulley added to the HG's usefulness as an agricultural workhorse, but it found many uses, included arctic exploration.

Right: Oliver took over Cletrac in 1944, hence the badge. They carried on making the HG in updated form until 1957.

Wide 68in tread on this HG, plus lots of track options, made it suitable for row-crop work.

Cletrac HG—the farmer's crawler.

Above: Rubber tracks are common now, and were actually an option on the HG. A few were sold with these, but they proved unreliable.

Left: Canadian plough maker Cockshutt also built its own tractors from 1945 to the early '60s. They were neat, good looking little machines.

Right: First Cockshutt was this one, the 30. The company pioneered a live (that is, continuously running) PTO, and live hydraulics.

Below: Smallest Cockshutt was the 20, with 20 drawbar hp, and 26hp at the belt, though the five-plough 50 was powered by a 273ci Buda six-cylinder engine.

Above: Cockshutt with unusual front-mounted harvester.

Above and left: Deutz of Germany was a true pioneer, both of tractors and the diesel engine. Its first tractor, a motor plough, was launched in 1907 and was considered quite advanced for its time. Its first diesel tractor, the MTZ222, arrived in 1926, a good quarter-century before the big US manufacturers began to run with the idea. Deutz produced a whole range of diesel tractors from the early 1930s. The Stahlschepper ("Iron Tractor") F1M, F2M, and F3M were powered by single-, twin-, and three-cylinder diesels respectively —the largest produced 50hp and was started by compressed air. So successful were these motors that Deutz sold them to other tractor manufacturers.

Below: Despite their decidedly European nature, Deutz tractors were among the first European machines to be exported to North America.

Right: This is a 1963 D25S, with twin-cylinder air-cooled diesel of 104ci (1.7 liter). For American farmers, the Deutz name is best known when linked with Allis from 1985.

Left: Fiat 315. Fiat of Italy has a strong connection with the US tractor market—its machines have been sold under both Allis-Chalmers and White badges in the past. The company has been building tractors for over 80 years.

Right: Fiat took over Ford New Holland in 1991, ensuring its access to North American markets. Now it retains a controlling interest in the giant CNH Global, which includes Case IH. This 680DT four-wheel-drive is a typical Fiat of the late 1970s.

Left: Fiat made use of all these connections to expand its own range—the Fiat G series of the late 1990s, for example, was nothing more than a repainted, rebadged Ford 70-series. But this 88-94 was Fiat born and bred—a mid-range machine that was an update of the 93 series.

Classic Profile

Huber isn't one of the better known American tractor manufacturers—in fact, outside enthusiast circles it's doubtful that anyone would recognize it at all. But in fact the company was in the tractor business for 40 years, longer than some of the more famous names of today.

Like many of the pioneer tractor makers, Huber, of Marion, Ohio, was already well established in the agricultural machinery line. It built steam traction engines, but lost no time experimenting with the new internal combustion technology. Its first IC machine was built as early as 1898, though this was not purpose designed—Huber simply bought a proprietory Van Duzen engine and bolted it into one of its existing steam engine chassis.

It was such a success that 30 units were built over the first year, and Huber took over Van Duzen to ensure its supply of engines. In fact, Huber was never to build its own IC engines, relying on well known makers like MidWest, Waukesha, and Stearns to supply them instead.

Smaller, purpose-designed tractors followed, like the 15-30, Huber's mid-range machine that used a 382ci MidWest four-cylinder engine. Huber was fond of this layout, and in fact every tractor it supplied for testing at Nebraska had four cylinders.

SPECIFICATIONS

Engine	382ci (6.1 liter)
Bore x stroke	4.5 x 6.0in
Transmission	2-speed
Horsepower	27/47hp
Fuel	9.23hp/hr per gallon
Weight	6,090lb (2,768kg)

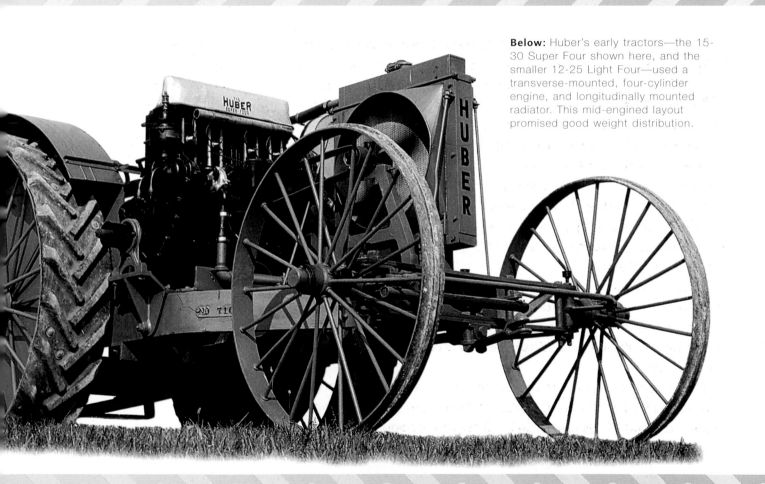

Below: Huber's early tractors—the 15-30 Super Four shown here, and the smaller 12-25 Light Four—used a transverse-mounted, four-cylinder engine, and longitudinally mounted radiator. This mid-engined layout promised good weight distribution.

Classic Profile

Below: From 1926, Huber tractors adopted a more conventional layout, with enclosed longitudinally mounted four-cylinder engines.

The biggest four it made was the 40-62, which Nebraska tested in 1927. Huber was remarkably modest in its power claims—the tractor had started life as a 25-50, until Nebraska found it produced 40% more brake horsepower than Huber claimed, and 60% more at the drawbar! The engine that performed this feat was a 617ci Stearns, rated at 1,100rpm.

Sadly, not all the Hubers tested at Nebraska performed so well. The last two—an LC and streamlined B in late 1937—both had problems. The LC slipped out of gear several times, its fuel tank leaked and the oil pressure gauge line broke. Maybe it was an omen, for there is no evidence that Huber survived World War Two as a maker of tractors.

Left: By the standards of the day, Huber was exceedingly modest in its power claims. Just as the big 25-50 turned out to be a 40-62 so this "15-30" produced 27hp at the drawbar and no less than 47hp at the belt, under test at Nebraska!

Below: Like every other Huber tractor engine, this one had four cylinders and wasn't made by Huber at all. The 382ci valve-in-head unit was bought in from MidWest, with Kingston Model L carburetor and LD4 magneto.

Henry Ford is usually given credit for popularizing the small, lightweight tractor in North America. The key word is "popularizing"— Henry wasn't the first person to sell a small tractor in the USA, just the first to make one cheap and practical enough for most farmers to buy.

Take the Bull Tractor Company. Its Little Bull was an exceedingly light and simple tractor, on sale from 1913. Weighing a little over 3,000lb when many contemporaries weighed around 20,000lb, it must have seemed absurd to those used to bigger machines. It was powered by a 230ci opposed twin-cylinder engine, and had three wheels, just one of which was driven to save the weight and cost of a differential. The direct gear drive gave only one speed, though the Little Bull did have the sophistication of a car-type radiator.

Little Bull should have been the ideal light and simple tractor, and at first, at the low price of $335, it sold very well. But there were drawbacks. With engine, radiator, fuel tank, and drive gear all mounted on the right hand side, the Little Bull was quite unstable. Despite a device to keep it level when the drive wheel was in the furrow, it was prone to tipping over. All the mechanics were exposed, and prone to rapid wear. Finally, single-wheel-drive and just five horsepower at the drawbar simply weren't enough for the rigors of field work.

Although around 5,000 Little Bulls were sold fairly quickly, word soon spread that it wasn't up to the job. Within a couple of years, the Little Bull had been withdrawn.

SPECIFICATIONS

Engine	230ci (3.7 liter)
Transmission	1-speed
Horsepower	5/12hp
Weight	3,200lb (1,455kg)
Numbers built	c5,000
Price new	$335

Right: Almost every moving part of the Little Bull was exposed to the wind and weather, dust and dirt, which did little for the longevity of the machine.

Left: Large gear wheel provided direct drive between engine and one rear wheel—cheap, but limited traction.

Classic Profile

Right: This step device was a means of adjusting the height of the non-driving rear wheel, to keep the tractor level when the other rear wheel was in the furrow.

Left: Lubrication came courtesy of a Detroit lubricator, and the twin-cylinder engine apparently held up well if treated gently.

Right: An automotive-style radiator was an unexpected piece of sophistication on the Little Bull, whose specification was otherwise pared to the cost cutting minimum.

Left: The Little Bull's driver could not see which way the front wheel was pointing, so a large arrow did the job for him.

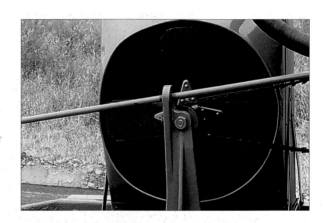

Left: The Kingston carburetor was a simple device, which summed up the Little Bull philosophy. Unfortunately, suspect stability and inadequate power killed it off within two years.

Above: The McCormick-Deering name would eventually be dropped.

Above: Badged a McCormick-Deering—built by International, in other words—this is the T40 TracTractor, which in this form was sold between 1932 and '39. This is one of the first built.

Right: The T40's six-cylinder 279ci engine came straight out of the W40 wheeled tractor, giving 43 brake horsepower. Five forward ratios gave speeds of 1.75 to 4.0mph.

Below: The gasoline powered T40 weighed in at 10,790lb, so it was no Caterpillar 2 Ton style light weight. In crawler terms, this was mid-range.

Right: International coined the catchy "TracTractor" brand name— Caterpillar executives must have wished they'd dreamt up that one!

Above: Other manufacturers soon followed in Caterpillar's rubber-tracked footsteps. This is a Case IH Steiger 9380, with Trac Trac system.

Right: One of the old school. Steiger made its name with conventional-wheeled, articulated tractors like this Tiger. The company is now part of CNH Global.

Right: White 4-150, in the company color scheme. This combined two Oliver two-wheel-drive power trains with a Cummins V8 diesel. It worked—a 175hp low profile articulated tractor with adjustable wheel tread. But White, which had taken over so many tractor makers, was itself bought by an investment company in 1980. That failed, as did the next owner. From 1993, AGCO provided a more stable home.

Above: "Field Boss" seems like the right sort of name for this White 4-225. It's an '83 model, powered by a Caterpillar V8 turbo diesel of 636ci (10.4 liter).

Right: White also sold a whole range of two-wheel-drive tractors in the 1970s and '80s. The smaller ones were imported from Iseki of Japan and Fiat of Italy, but larger machines like this one, up to the Caterpillar powered 2-180, with its Caterpillar V8, were built by White itself.

INDEX

ACKNOWLEDGEMENTS & PICTURE CREDITS

The publishers would like to thank the photographers, manufacturing companies, and private collectors who have cooperated in the production of this book either by supplying material from their archives or by allowing access to their collections of vehicles and memorabilia for photography. The black and white pictures were supplied by the Ford Motor Company and Caterpillar Inc, and are reproduced with their kind permission.

The majority of the color photographs in this book were taken by Andrew Morland. The commissioned photographs of Classic Profiles were taken by Neil Sutherland, with the exception of the pictures on pages 48–51, 122–5, 130–1, 246–7, 354–7, 368–9, and 384–7, which were supplied by Andrew Morland.